Springer Series in Optical Sciences Volume 24
Edited by Theodor Tamir

Springer Series in Optical Sciences

1 **Solid-State Laser Engineering**
By. W. Koechner

2 **Table of Laser Lines in Gases and Vapors**
3rd Edition
By R. Beck, W. Englisch, and K. Gürs

3 **Tunable Lasers and Applications**
Editors: A. Mooradian, T. Jaeger, and P. Stokseth

4 **Nonlinear Laser Spectroscopy**
By V. S. Letokhov and V. P. Chebotayev

5 **Optics and Lasers**
An Engineering Physics Approach
By M. Young

6 **Photoelectron Statistics**
With Applications to Spectroscopy and Optical Communication
By B. Saleh

7 **Laser Spectroscopy III**
Editors: J. L. Hall and J. L. Carlsten

8 **Frontiers in Visual Science**
Editors: S. J. Cool and E. J. Smith III

9 **High-Power Lasers and Applications**
2nd Printing
Editors: K.-L. Kompa and H. Walther

10 **Detection of Optical and Infrared Radiation**
2nd Printing
By R. H. Kingston

11 **Matrix Theory of Photoelasticity**
By P. S. Theocaris and E. E. Gdoutos

12 **The Monte Carlo Method in Atmospheric Optics**
By G. I. Marchuk, G. A. Mikhailov, M. A. Nazaraliev, R. A. Darbinian, B. A. Kargin, and B. S. Elepov

13 **Physiological Optics**
By Y. Le Grand and S. G. El Hage

14 **Laser Crystals** Physics and Properties
By A. A. Kaminskii

15 **X-Ray Spectroscopy**
By B. K. Agarwal

16 **Holographic Interferometry**
From the Scope of Deformation Analysis of Opaque Bodies
By W. Schumann and M. Dubas

17 **Nonlinear Optics of Free Atoms and Molecules**
By D. C. Hanna, M. A. Yuratich, D. Cotter

18 **Holography in Medicine and Biology**
Editor: G. von Bally

19 **Color Theory and Its Application in Art and Design**
By G. A. Agoston

20 **Interferometry by Holography**
By Yu. I. Ostrovsky, M. M. Butusov, G. V. Ostrovskaya

21 **Laser Spectroscopy IV**
Editors: H. Walther, K. W. Rothe

22 **Lasers in Photomedicine and Photobiology**
Editors: R. Pratesi and C. A. Sacchi

23 **Vertebrate Photoreceptor Optics**
Editors: J. M. Enoch and F. L. Tobey, Jr.

24 **Optical Fiber Systems and Their Components**
An Introduction
By A. B. Sharma, S. J. Halme, and M. M. Butusov

25 **High Peak Power Nd: Glass Laser Systems**
By D. C. Brown

26 **Lasers and Applications**
Editors: A. Mooradian, W. O. N. Guimares, and C. T. Lin

27 **Color Measurement**
Theme and Variations
By D. L. MacAdam

28 **Modular Optical Design**
By O. N. Stavroudis

A.B. Sharma S. J. Halme M. M. Butusov

Optical Fiber Systems and Their Components

An Introduction

With 125 Figures

Springer-Verlag Berlin Heidelberg New York 1981

Dr. Awuashilal B. Sharma
Professor Seppo J. Halme

Communication Laboratory,
Helsinki University of Technology,
02150 Espoo 15, Finland

Professor Mikhail M. Butusov

Leningrad Polytechnical Institute,
195251 Leningrad, USSR

ISBN 3-540-10437-2 Springer-Verlag Berlin Heidelberg New York
ISBN 0-387-10437-2 Springer-Verlag New York Heidelberg Berlin

Library of Congress Cataloging in Publication Data. Sharma, Awuashilal B. 1948-. Optical fiber systems and their components. (Springer series in optical sciences ; v. 24) Bibliography: p. Includes index. 1. Fiber optics. 2. Optical communications. I. Halme, Seppo J., joint author. II. Butusov, Mikhail Mikhailovich, joint author. III. Title. TA1800.S47 621.36'92 80-25865

Offset printing: Beltz Offsetdruck, Hemsbach/Bergstr. Bookbinding: J. Schäffer oHG, Grünstadt.
2153/3130-543210

Preface

This book is intended to be an introductory text for engineers and physicists who are likely to be involved in the area of optical fiber communications. Its purpose is to provide the student with an explanatory text that can also be used for "self-study". Thus, key theoretical results have been rather thoroughly derived, and detailed explanations have been given wherever certain steps have been excluded. Some of the derivations are in new form, which the reader will hopefully find stimulating. In addition, some of the experimental and theoretical results are based on the research of the authors, and they are published here for the first time. However, references are given for all those cases involving equivalent results obtained by others.

Although a large number of monographs are available for the specialist or the knowledgeable scientist, most of these are inadequate for teaching purposes. This aspect served as a major motivation for writing a book that explains the basic phenomena and techniques. The required material was partly developed in earlier courses on integrated optics and optical fiber communications, and partly resulted from the authors' close cooperation with industry. To assess the suitability of the material, the manuscript of the book was used with encouraging results for a graduate course (spring semester, 1980) at the Communications Laboratory of the Helsinki University of Technology. In the future, the book will also be used for teaching undergraduate courses. The prerequisite is the usual mathematical and physical background of advanced undergraduates in electrical engineering and physics.

The proposal for writing this book was made by M.M. Butusov of the M.I. Kalinin Polytechnic Institute of Leningrad, while on a research fellowship at the Communications Laboratory in 1979. The manuscript was prepared collectively, with one of us (A.B. Sharma) responsible also for editing and coordinating the material. The work has been considerably improved by the critical reviews of Prof. T. Tamir, our supervising editor. We are also grateful to Messrs. M. Hall, K. Horko, E.J.R. Hubach, J. Saijonmaa, and

S. Törmälä of the Communications Laboratory, as well as to all the students and industrial personnel who participated in our spring semester course, for pointing out errors and inconsistencies in the text. Mr. Saijonmaa was also the source of many of the numerical results for field computations.

In addition, we would like to thank the General Direction of Posts and Telecommunications (Posti- ja lennätinlaitos), the Ministry of Education (Opetusministeriö), and the Foundation for the Advancement of Technology (Tekniikan Edistämissäätiö) for their financial support.

Finally, it is with pleasure that we acknowledge the painstaking and excellent work of Mrs. Aila Halme, who typed the manuscript, and Mr. Harri Rimmes, who prepared the line drawings.

Helsinki, Finland,
October, 1980

A.B. Sharma, S.J. Halme,
M.M. Butusov

Contents

1. Introduction

1.1 Historical Background

In the broadest sense, communication by optical means has been intermittently practiced since the times of antiquity, by taking natural advantage of the sense of vision. Smoke signals during the day, and fires at night, have been used in many cultures to transmit warnings about enemy movements, the outcome of battles, etc. For example, it is known that fire signals were extensively used by the armies of the Roman empire [1.1].

The turning point from such primitive systems occured in 1791, with C. Chappe's invention of the optical telegraph. The success of Chappe's system is demonstrated by both its widespread use in France (for more than 60 years), and the large number of equivalent systems that sprang up in Europe. For example, it is known that A. Edelcrantz developed a binary-coded-decimal system in 1794, which was later used across the Åland Sea [1.1]. In any case, the optical telegraph era demonstrated the benefits of even rather simple optical systems, and also underscored the severe dependence of such atomospheric techniques upon weather conditions. In fact, the vagaries of the weather were probably an important reason for the success of Marconi's "wire-less" telegraph, which was more or less independent of weather and visibility.

The next landmark in the history of optical communications can probably be attributed to A.G. Bell who experimented with the optical transmission of speech [1.2]. Unfortunately, at that point, the idea of optical communications was discontinued because of both the necessity of an entirely new electro-optical technology, and the rapid development of radio systems. The idea lay dormant for a number of decades and re-emerged only after gas and solid-state lasers became available in the early 1960's. However, the transmission medium was still the atmosphere and, as in the 18th and 19th centuries, the weather was once again found to be a limiting factor. An optical waveguide was obviously needed to ensure reliable transmission.

The concept of a dielectric waveguide was certainly not unfamiliar, as it had been analyzed in 1910 by HONDROS and DEBYE [1.3] who were interested in the propagation of radio waves along dielectric wires. Many decades later, KAPANY [1.4] published his work in 1967 on dielectric waveguides for imaging applications at optical frequencies. However, the loss in the available materials, of which glasses were the most interesting, was far too high for application in communications systems. The reason for the high loss was recognized by KAO and HOCKHAM in 1966 [1.5], when they speculated that sufficiently low-loss fibers could be produced by the use of ultra-pure glasses. They were proved correct some four years later, when KAPRON et al. [1.6] reported a quartz fiber with an attenuation of about 20 dB/km. This was followed by approximately a decade of explosive development in which attenuation figures went down and bandwidths went up. Simultaneously, the performance of semiconductor light sources was dramatically improved.

Today, we can say that the feasibility of optical fiber communications has been clearly demonstrated by the large number of trial systems that can be found in most of the industrialized countries of the world. These systems have not only confirmed the inherent advantages of optical fibers (immunity to electro-magnetic interference, light weight, abundance of raw materials, etc.), but have also clearly shown that inter-repeater distances can be long even in comparison to those of co-axial cable systems. Indeed, given the intense involvement of most major communications organizations, it is felt that introductory courses on optical fiber systems should now be a part of the curriculum of future generations of scientists and engineers who are likely to be involved in this exciting and challenging area. This book is intended to form the basis of such a course, and the material has been adapted accordingly. For basic papers and other texts covering similar material, the reader should consult [1.7-12].

1.2 Outline of Contents

We start in Chap.2 with a semi-qualitative review of the physics describing the generation, modulation, and detection of light. The discussion of generation and detection is restricted to the phenomena essential for the understanding of the principles behind the operation of light-emitting diodes, solid-state lasers, and semiconductor diode detectors. In the section on modulation, the electro-optical, acousto-optical, and magneto-optical effects are discussed. We also indulge in some minor speculation about some other effects that might have future applications.

In Chap.3, we first review the essentials of total internal reflection at a plane interface (Sect.3.1), and then at curved, blurred, and periodically modulated interfaces (Sect.3.2). The treatment is intended to be explanatory rather than rigorous, and plausibility arguments are often used. The purpose is to establish a physical basis for an intuitive understanding of wave behaviour in optical waveguides.

The information of Sects.3.1 and 3.2 is then used to introduce the reader to the planar step-index waveguide in Sect.3.3, in which the reader is also exposed to the concept of the discrete modes of a resonator. This is done using simple arguments and a convincing graphical technique. Once again, the motivation is to provide an intuitive understanding.

The level of mathematical difficulty is somewhat increased in Sect.3.4, in which the solution of Maxwell's equations for a planar graded-index waveguide is considered by using the WKB method and Airy functions. This is in preparation for the analysis of graded-index fibers so that rather complete derivations are provided, and detailed explanations are included. However, before this derivation, a ray analysis is given in order to show the manner in which radiation is confined within such a structure.

In Sect.3.5, after a qualitative discussion of the modes of a step-index fiber, the rest of the section is almost exclusively devoted to the basic technique used for the analysis and computation of the impulse response of multi-mode fibers. A detailed derivation of the eigenvalue equation is given, and is shown to reduce into its simplified form in the "weakly guiding" approximation. The purpose of this section is to clearly explain the steps involved in the analysis of the fields of a fiber, and to lay a firm foundation for the more complex procedures of Sect.3.6, in which graded-index fibers are treated.

Section 3.6 is, in fact, the culmination of all previous sections of Chap.3, and includes both the three-dimensional counterpart of Sect.3.4, and the generalization (within the WKB-Airy function approximation) of Sect.3.5. After derivation of the eigenvalue integral, and its exact solution for step-index and parabolic profiles, analytical solutions of the impulse response are presented using the α-profile description. Once again, the derivations are detailed, with abundant explanations.

The chapter concludes with a brief discussion of coherence effects (Sect. 3.7), in order to provide a basis for the discussion of the important and troublesome phenomenon of "modal noise" in the analog system of Sect.6.2.

In Chap.4, we review the actual components that go towards the formation of an optical fiber system, and discuss the means available for inter-connecting these components. The principles behind the fabrication of fibers are discussed in Sect.4.1, while optical sources are considered in Sect.4.2. In the latter, the stress is on semiconductor double-heterojunction lasers, but a short review of fiber lasers is also given. For the sake of completeness, thin-film modulators are shortly reviewed in Sect.4.3.

The structure and performance of PIN and avalanche photodiodes are considered in Sect.4.4, in order to lay the foundation for the noise analysis of Sect. 6.1, and the chapter concludes with a treatment of the coupling problem. The stress has been placed on source-to-fiber coupling, and detailed derivations are given. The question of whether coupling can be improved by the use of optical elements has been treated by the use of Liouville's theorem. The "optimum" efficiencies yielded thereby have been compared with some practically attainable values.

Chapter 5 has been devoted to the presentation of the fundamentals of fiber measurements. The parameters of interest have been restricted to attenuation (Sect.5.2), impulse response (Sect.5.3), frequency response (Sect.5.4), and the refractive-index profile (Sect.5.5). The last of these has been further restricted to a consideration of the near-field scanning method, for which a detailed analysis of the influence of leaky modes is given. The chapter includes a discussion of the many problems associated with the multi-mode nature of the fibers under consideration.

We conclude the book with Chap.6, in which we give a thorough analysis of the design and performance of optical fiber systems. The stress is on digital systems (Sect.6.1), and both attenuation-limited and dispersion-limited links are discussed. The analysis is based on the concept of equivalent bandwidths, and has been clarified with the aid of numerous examples. The purpose is to provide a unified approach that is generally applicable in the different design situations encountered in practice. With some minor modifications, the analysis has also been used for the evaluation of the analog systems briefly discussed in Sect.6.2.

For the sake of completeness, Sects.6.3 and 6.4 have been devoted to a short review of instrumentation and data systems, and fiber sensors, respectively. Finally, in Sect.6.5, we briefly introduce the reader to the technique of wavelength-division multiplexing, whereby the capacity of time-division-multiplexed systems can be further increased.

In ending this section, we would like to remind the reader that, throughout this book, we adhere to the SI system of units.

2. Generation, Modulation, and Detection

An optical fiber communications system, in much the same way as its more conventional relatives, requires an optical source which provides the information "carrier", a means for modulating a suitable parameter of the carrier, a channel for propagation (the fiber), and a scheme for detecting the carrier at the receiver end. When sources such as light-emitting or laser diodes are used, modulation is simply a matter of suitably varying the drive current. On the other hand, with sources such as fiber or solid state (e.g. Nd-YAG) lasers, a modulator external to the active material is normally required. At the receiver end, the modulation is normally detected by using photosensitive p-n junctions of various types. It is then obvious that, besides propagation effects in the fiber (to be treated in Chapt.3), we are also interested in sources, modulators, and detectors.

In this chapter our main purpose is to review the physical ideas pertinent to the understanding of these elements, whereas details of the components themselves are deferred to Chapter 4. Moreover, we assume that the reader has, in fact, been previously exposed to solid-state physics, and that he or she only needs to be reminded of the principal features.

2.1 Generation of Light in Solids

We start with a consideration of the energy band structure which can decisively influence the optical and electronic properties of a crystalline solid. The band structure is usually presented in the form of E-k relationships obtained by solving the Schrödinger wave equation for the single-electron problem. Thus, we consider the quantum behaviour of an electron within a perfectly periodical potential. We can think of such a potential field to be due to the ion "cores" situated at lattice sites. In addition, we must also include an average "smeared" potential distribution due to all the free electrons that are associated with the crystal (Fig.2.1). Denoting the total potential field by $u(\underline{r})$, the Schrödinger's wave equation takes the form [2.1]

$$\left[-\frac{\hbar^2}{2m}\nabla^2 + u(\underline{r})\right]\psi_k(\underline{r}) = E_k\psi_k(\underline{r}) \quad , \tag{2.1}$$

where $\hbar = h/(2\pi)$, ψ is the wave function of the electron, m is its mass, and h is Planck's constant. We look for solutions that are Bloch functions of the form:

$$\psi_k(\underline{r}) = \phi_k(\underline{r})\exp(j\underline{k}\cdot\underline{r}) \quad . \tag{2.2}$$

In other words, we assume that the wavefunction is a plane wave which is periodic in $\underline{r}$ (with the periodicity of the direct lattice) and has an amplitude function $\phi_k(\underline{r})$. In fact, the amplitude function accounts for the influence of the periodic potential $u(\underline{r})$ on the free electron wavefunction $\exp(j\underline{k}\cdot\underline{r})$. Satisfaction of boundary conditions then gives the dependence of electron energy E on its momentum k (Fig.2.2). The dashed parabolic curve represents the free electron case when $E = \hbar^2k^2/(2m)$ and there are no restraints on the possible values of energy. The influence of the lattice is to break the curve at values of $k = n\pi/a$, where "a" is the lattice constant (Fig.2.1). As will be seen in Chap.3, an analogous situation occurs in an optical resonator which forces a plane wave to propagate at certain allowed angles.

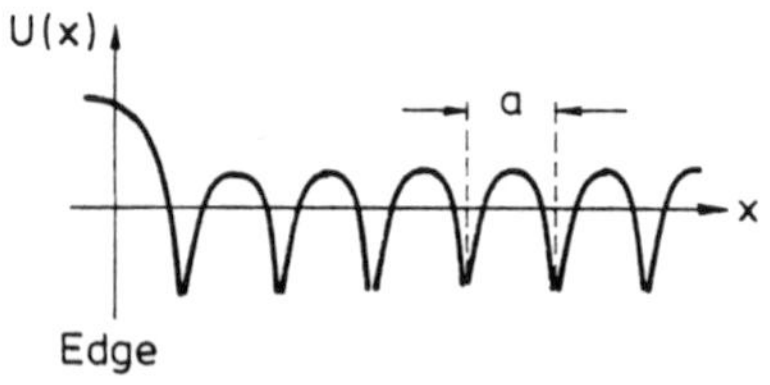

Fig.2.1. Schematic representation of the potential distribution in a crystal

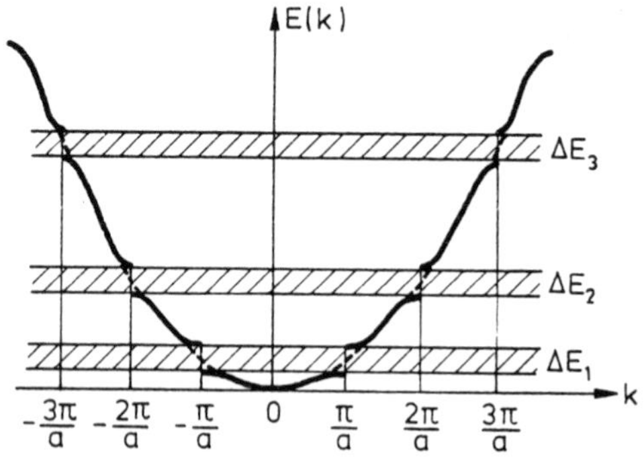

Fig.2.2. Dependence of electron energy on its momentum

A physical interpretation of the forbidden energy values can be based on the internal *Bragg diffraction* of the electron wave at the periodic (one-dimensional in Figs.2.1 and 2.2) structure. To support this interpretation, we must look at the de Broglie wavelength of the electron

$$\lambda = \frac{2\pi}{k} = \frac{h}{mv} \quad . \tag{2.3}$$

Comparing (2.3) with the k-values at which forbidden energy gaps appear, we easily conclude that the well-known Bragg condition is satisfied at these values, i.e.

$$\lambda_{Br} = \frac{2\pi}{k_{Br}} = \frac{2a}{n} \quad ,$$

$$\text{or} \quad a = n \frac{\lambda_{Br}}{2} \quad . \tag{2.4}$$

Recalling now the formula of the group velocity v_g of the "wave packet" (the superposition of plane waves with close k-values)

$$v_g = \frac{d\omega}{dk} = \frac{1}{\hbar} \frac{\partial E}{\partial k} \quad , \tag{2.5}$$

we see from Fig.2.2 that, as we approach the forbidden region, the group velocity tends to zero, namely

$$v_g \Big/_{k = \frac{n\pi}{a}} = 0 \quad , \tag{2.6}$$

and the wave-packet with this momentum cannot propagate. Confining the energy diagram to the first Brillouin zone, i.e. $-\pi \leq ka \leq \pi$, we obtain Fig.2.3, which describes the energy situation in the so called reduced-zone representation. Such confinement is justified by the similarity of all unit cells in the crystal [2.1]. Based on the above, and using Pauli's exclusion principle, we arrive at the next significant conclusion: within each energy band, there are 2N levels which can be occupied by electrons. Some of these levels may be entirely filled, others partially filled, yet others completely empty, depending on the number of electrons provided by atoms. Furthermore, because electrons

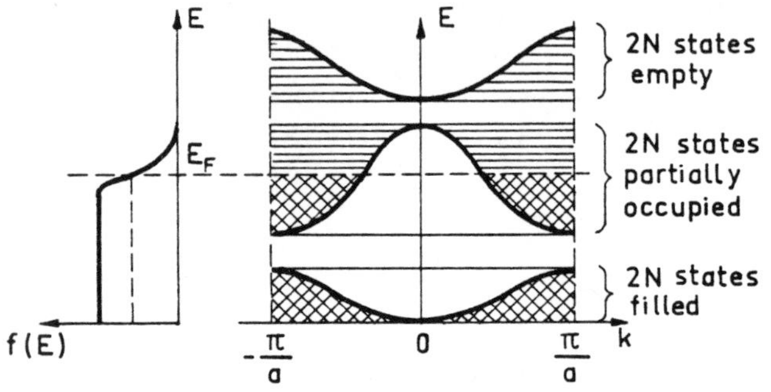

Fig.2.3. Reduced zone representation of the band structure

obey Fermi statistics, only those near the Fermi level E_F (see Fig.2.3) can move under the influence of an external electric field. The position of the Fermi level is therefore a factor of decisive importance in determining the behaviour of the crystal.

When the highest band is completely filled with electrons, the Fermi level coincides with the center of the forbidden gap, and the electrons are not free to move upon the application of an electric field. This case is illustrated by Fig.2.4a, and is relevant to intrinsic semiconductors and insulators, which only differ in the width of their forbidden gap. (ΔE is of course much larger in insulators). On the other hand, when the last band is only partially filled, as in metals, the whole top region of the Fermi distribution can be used for recruiting carriers for conduction, as shown in Fig.2.4b.

Intrinsic semiconductors with sufficiently narrow forbidden regions may enter into the same situation if some perturbation, such as heat or light, gives the electrons sufficient energy to "jump the gap" and become conduction

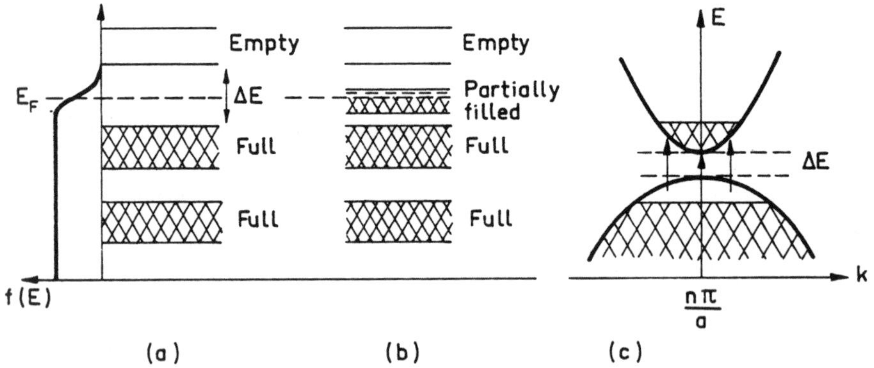

Fig.2.4a-c. Electron densities in (a) an insulator or semiconductor, (b) a conductor, and (c) a conducting semiconductor

electrons, as illustrated in Fig.2.4c. The formerly empty band now becomes the conduction band, while the electrons' previous "home" is called the valence band. In leaving the valence band, these electrons produce empty levels, so that conduction can now occur in both bands.

At this point, we digress slightly and note that the effective mass of an electron obeys the law [2.1]

$$m^* = h^2/(d^2E/dk^2) \quad . \tag{2.7}$$

The immediate implication of (2.7) for the valence band (for which the E-k curve is concave downwards) is that electrons at the top of the valence bands have negative mass, or that they experience acceleration in a direction opposite to that of "normal" free electrons. A *negative particle with negative mass* would be accelerated in the same direction as a *positive particle with positive mass*. The situation in the nearly filled valence band can hence be regarded as one involving positive particles, commonly referred to as "holes". Thus, excitation of electrons from the valence band to the conduction band is accompanied by the excitation of an equal number of holes in the valence band.

We must now remind ourselves that our conclusions so far are based on the one-dimensional, over-simplified approach. A more accurate analysis would have to account both for the three-dimensional situation, and for the possibility of having overlap amongst different bands. The energy-band structures of the best known semiconductors (Ge, Si, GaAs) are shown in Fig.2.5. With the help of these, we can reach two further conclusions.

1) We can have two valence bands with coincident peaks but different curvatures. Thus, in accordance with (2.7), "light" and "heavy" holes can exist.
2) The nearest energy valley of the conduction band is not necessarily set above the peak of the valence band (Si and Ge), implying that indirect transitions (accompanied by momentum transfer, presumably to the lattice vibrations) are also incorporated in the electron excitation process.

Given such direct and indirect band-gap materials, let us next consider the manner in which electrons can be excited from the valence to the conduction band. First, we note that in the type of intrinsic semiconductor we are now considering, the excitation energy will be mainly thermal or optical. Consider, for example, that a photon of energy $h\nu = \Delta E$ is absorbed and causes the excitation of an electron. In the process, momentum must be conserved. In a direct band gap material (Fig.2.5c) we have no problem, because there is no momentum difference between the top of the valence band and the bottom of the conduction band. However, in an indirect material, a change in momentum is required. As the momentum of a photon is negligible, the momentum change has to be caused by another agency. This is normally attributed to the creation or annihilation of a phonon with a suitable momentum vector because the associated phonon energy is characteristically small.

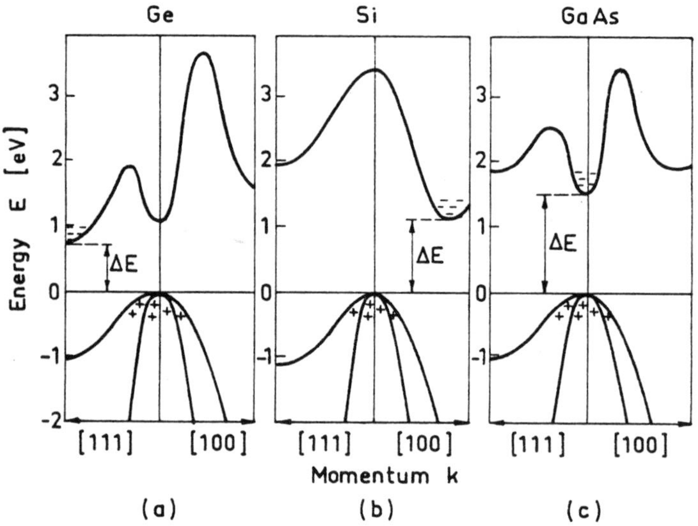

Fig.2.5a-c. Energy band structure of some common semiconductors (a) germanium, (b) silicon, and (c) gallium arsenide

In this section, the phenomenon of interest to us is the opposite of the above process - the recombination of electron-hole pairs after excitation. Once again, a crucial distinction exists between direct and indirect band-gap semiconductors. For the former types, the electron can simply go back to the valence band and re-combine, giving up its excess energy in the form of a photon of energy $h\nu = \Delta E$. For indirect band gaps, the generation or absorption of a phonon is required, as for the excitation process. However, such a simultaneous interaction of a photon and a phonon inevitably reduces the probability of the event, and the electron will tend to dwell in the conduction band for relatively long times. Moreover, there may also be competing non-radiative transfers in the material due to defects and impurities. The overall result is that the probability of radiative band-to-band recombinations in indirect semiconductors is significantly lower than in direct band-gap types.

In practical semiconductor light sources, the process of electron-hole pair "generation" is normally that of carrier injection, as in p-n junctions. In order to describe this process, we must first look at the effect of dopants on the energy band structure of the intrinsic material. The first significant difference is that impurity energy levels are created [2.2,3], as shown in Figs.2.6a and c for "donor" and "acceptor" impurities, respectively. (The names arise from the fact that "donors" give up electrons to the conduction band, while "acceptors" receive electrons from the valence band). As the doping levels are increased, the donor and acceptor levels broaden out into impurity bands, until they eventually overlap (in degenerate semiconductors) with the conduction (or valence) band, as shown in Figs.2.6b and d. The consequent change in the band gap can lead to significant changes in the optical

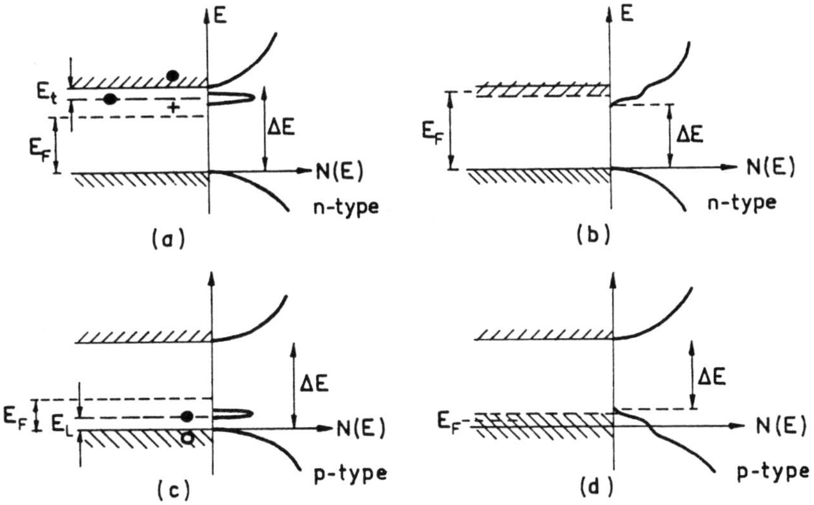

Fig.2.6a-d. The effect of impurities on the band structure of semiconductors: (a) lightly donor doped, (b) heavily donor doped, (c) lightly acceptor doped, and (d) heavily acceptor doped

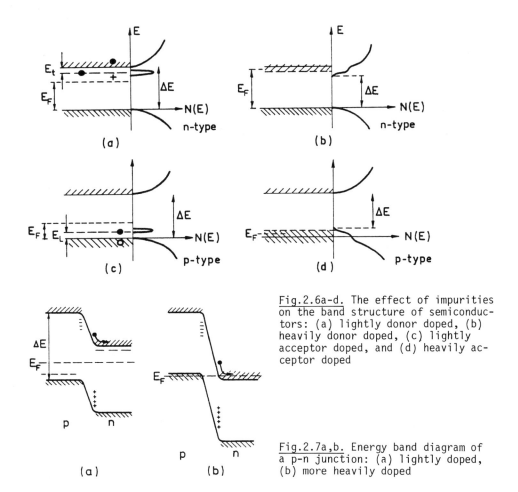

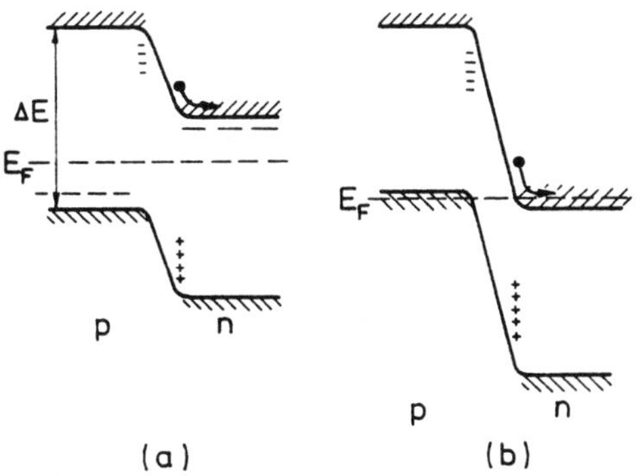

Fig.2.7a,b. Energy band diagram of a p-n junction: (a) lightly doped, (b) more heavily doped

properties. Moreover, the higher the doping concentration, the closer is the Fermi level to the dopant level, until ultimately it also merges with the conduction (or valence) band.

Now we turn to the previously mentioned p-n junction, which will be used for injection excitation and subsequent light generation. Such a junction is formed by the tight contact between donor doped (n-type) and acceptor doped (p-type) semiconductors. The resultant energy diagrams are shown in Figs.2.7a and b, for lightly and heavily doped semiconductors, respectively. The nature of the diagrams can be understood by considering that, in thermal equilibrium, the Fermi level must be the same on either side of the junction, thereby leading to a potential difference across the junction. A simple physical picture can also be obtained by bearing in mind that every donor atom, having lost an electron, becomes a "fixed" center of positive charge; similarly, every acceptor atom corresponds to a fixed negative charge. Charge neutrality is maintained by the free electrons and holes. If we form a junction of n- and p-type semiconductors, the electrons would tend to diffuse to the p-side while holes would go to the n-side. However, as soon as this happens, a positive (excess) charge is left behind on the n-side, and a negative one on the p-side. These together inhibit further diffusion, and create a "depletion" region within which there is a negligibly small number of free carriers. In any case, the existence of these positive and negative charge distributions produces a potential (voltage) gradient, defined by Poisson's equation [2.3]. Equivalently, we can say that the energy level structure assumes a graded form, as in Figs.2.7a and b.

If this potential gradient is lowered by the application of an external voltage, free carriers can once more begin to flow across the junction. In direct band-gap semiconductors with suitable levels and band gaps, band-to-band radiative recombinations can then occur. It is important to note that, at low temperatures, when centers within the forbidden band gap may be unoccupied, recombinations other than band-to-band can also produce "injection-luminescence", particularly with shallow traps. However, at room temperatures, such traps are easily occupied and can be neglected [2.4]. We conclude that, in direct band-gap materials at room temperatures, intrinsic electron-hole recombinations provide the most important mechanism for the generation of light.

So far, our discussions have been restricted to the usual spontaneous recombinations that form the basis for the operation of light emitting diodes. Another form of radiative recombination, which was theoretically predicted by EINSTEIN [2.5], is known as stimulated emission. This form of recombination, which has no classical analogs, occurs when a photon "triggers" the process that produces a second photon. The essential feature of stimulated emission is that the second photon assumes the optical properties (frequency, phase, polarization) of the triggering photon, as depicted in Fig.2.8. This property of coherence accounts for the narrow spectral widths of lasers and their capability of being able to produce contrast-interference pictures under a wide variety of conditions.

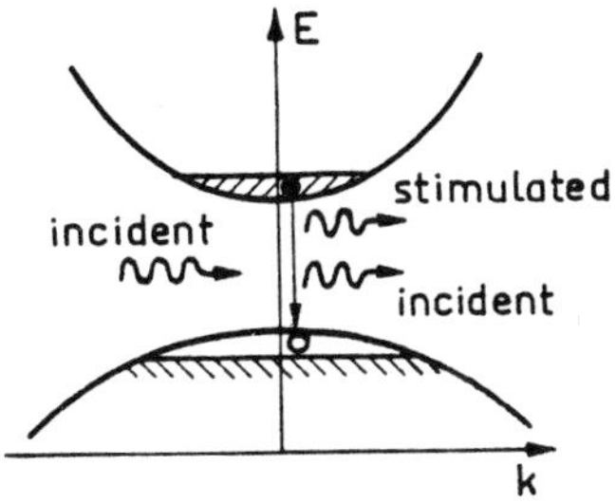

Fig.2.8. Stimulated emission due to an incident photon

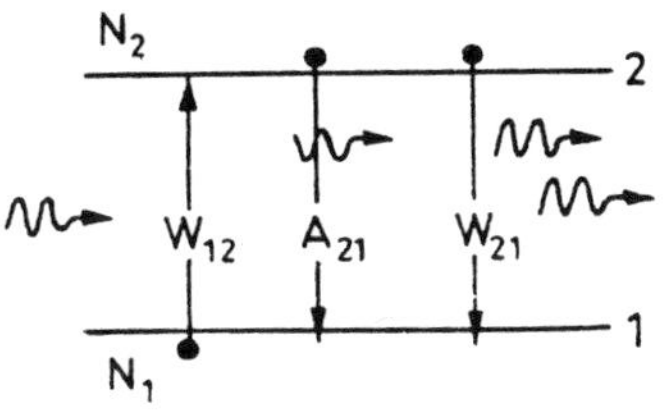

Fig.2.9. A simple two-level system showing excitation, spontaneous emission, and stimulated emission

Let us consider the competition between different electron transition processes in a simple two level system (Fig.2.9). We assume that there are three possible transitions: excitation (due to optical pumping power) at a rate W_{12} per atom, spontaneous recombination at a rate A_{21} per atom, and stimulated recombination at a rate W_{21} per atom. The rate equation for this case takes the form

$$\frac{dN_2}{dt} = W_{12}N_1 - (W_{21} + A_{21})N_2 \quad , \tag{2.8}$$

where N_1 and N_2 are the population densities of electrons in the lower and upper levels, respectively. In the steady state, $dN_2/dt = 0$, and (2.8) reduces to

$$\frac{N_2}{N_1} = \frac{W_{12}}{W_{21}+A_{21}} \quad . \tag{2.9}$$

It can be shown [2.6], that we must have *population inversion*, or the population density N_2 must exceed N_1 in order to achieve a process of amplification. In our simple two level system (implying a sparse atomic density as in gases), this cannot be achieved by optical pumping because excitation and stimulated emission are equally likely ($W_{12} = W_{21}$), while spontaneous emission always occurs ($A_{21} > 0$). However, the above consideration lays the foundation for considering the possibilities of population inversion in direct band-gap semiconductors.

As discussed earlier, semiconductor atoms are very tightly packed, causing their energy levels to overlap into bands. In such a situation, the possibility of population inversion can be illustrated with the aid of Fig.2.10. Thus, suppose that we have injected electrons into the conduction band up to the quasi-Fermi level E_{Fc}. Because an equal density of holes is generated, the states in the valence band, up to the level E_{Fv}, must be empty to preserve charge neutrality in the material. Thus, photons with energy greater than ΔE but less than $(E_{Fc} - E_{Fv})$ cannot be absorbed, because the conduction band states are already full up to E_{Fc}. However, these photons *can* produce stimulated emission and induce downward transitions. The situation we have considered applies at absolute zero temperature, in which case the sharp boundaries shown in Fig.2.10 exist. However, as the temperature is raised, a re-distribution of the electrons and holes occurs, and this smears the sharply defined carrier distribution. Nevertheless, the basic conditions for population inversion and stimulated emission remain.

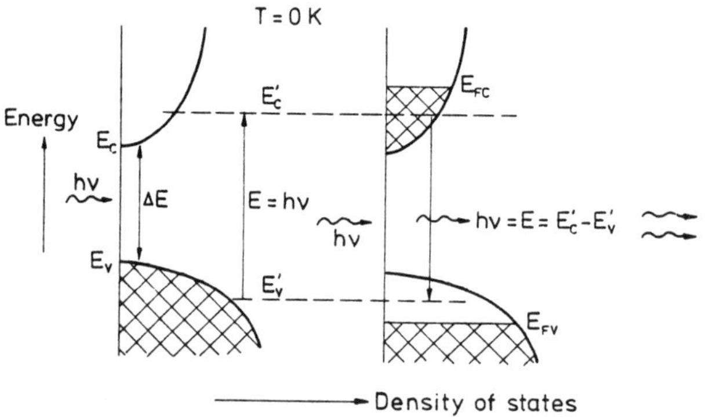

Fig.2.10. Illustration of population inversion in a semiconductor

As mentioned previously, population inversion cannot be achieved in a two-level system by optical pumping. Thus, in fiber and other solid state lasers (e.g. ruby), another approach is used. Consider the three-level system of Fig.2.11a for which the rate equation takes the form

$$\frac{dN_2}{dt} = (W_p + W_{12})N_1 - (W_{21} + A_{21})N_2 \quad , \tag{2.10}$$

where W_p is the pumping rate for excitation to the 3rd level (band), and the subsequent rate of non-radiative transitions from the 3rd to the 2nd level. If we again assume that $W_{12} = W_{21}$, to achieve population inversion ($N_2 > N_1$) we must have

$$W_p > A_{21} \quad . \tag{2.11}$$

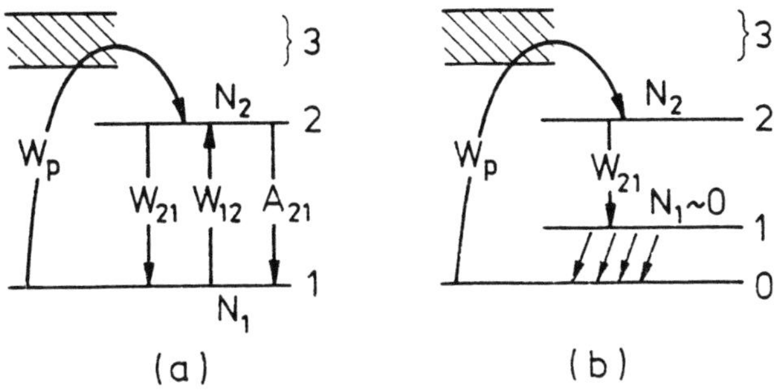

Fig.2.11a,b. Optical processes in (a) a three-level system, and (b) a four-level system

Hence, we must pump to level 2 faster than it is depleted by spontaneous emission. Of course, the rate W_p depends both on the energy density of the pump source and on the absorption properties of the material. The desirable conditions can then be summarized to be

1) an intense pumping source,
2) strong absorption of the pump light (implying a broad band for the 3rd level, and a suitable pump spectrum),
3) a metastable 2nd level to reduce A_{21}.

These conditions can be eased somewhat by inserting a fourth level into the structure, as illustrated in Fig.2.11b. The material should then be such that it allows fast depletion of the lower working level (1) to the ground state (0). Under these conditions, $N_1 \approx 0$, and population inversion in level 2 can be easily achieved. A second advantage is that such a material does not absorb its own radiation, because the lower level, responsible for optical transitions, is unpopulated. Such lasers are therefore more efficient in comparison to their three-level counterparts.

We have seen how an optical gain greater than unity can be achieved, and must now look at how a self-oscillating source can be made. As in the case of electrical oscillators, we require positive feedback, preferably resonant, in order to obtain a freely oscillating system. This feedback is typically achieved using a Fabry-Perot cavity, as illustrated in Fig.2.12a, which consists of two parallel mirrors, one of them only partially reflecting. Power can then be extracted from this latter one. In the case of semiconductors, the cavity is usually formed by cleaving along parallel planes, for example, the [110] plane in zinc-blende materials.

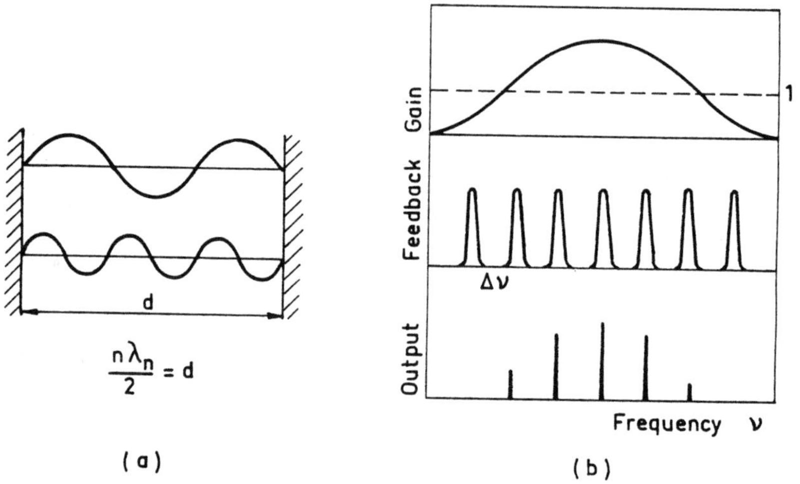

Fig.2.12. (a) A Fabry-Perot resonator. (b) Influence of the resonator on laser performance

It can be shown [2.6,7], that a resonator allots narrow spectral lines out of the radiation of the active medium. These lines, called the *longitudinal modes* , are separated by the free spectral range at which no feedback is obtained[1]

$$\Delta\lambda = \lambda^2/(2d) \quad , \quad \text{or} \tag{2.12a}$$

$$\Delta\nu = c/(2d) \quad . \tag{2.12b}$$

---

1 The physical meaning of (2.12) is, of course, that a standing wave pattern can only be established at selected wavelengths.

The effect of the optical cavity on laser performance, is illustrated in Fig.2.12b. The top curve shows the frequency dependence of the gain. Obviously, oscillation is only possible for gains greater than unity. The middle curve shows the reflectance peaks of the resonator, at which strong feed-back occurs. The bottom curve shows the final laser output, which, of course, must fulfill the conditions set by the two curves before it.

2.2 Modulation of Light

It is obvious from Sect.2.1 that the output optical power of semiconductor lasers and light-emitting diodes can be simply modulated by varying the injection current. However, the use of a modulator external to the source widens our choice and allows the use of both conventional (coherent and incoherent) sources, and newer types such as fiber-lasers. Moreover, in view of the future of ultra-broadband single mode fibers having low loss, it would be desirable to have modulation possibilities well into the microwave range. The use of an external modulator placed in a microwave circuit allows us to anticipate such bandwidth possibilities, particularly when we think in terms of thin film modulators [2.8-10]. This occurs because, with conventional "bulk" modulators, the device capacitance is rather high, but can be substantially reduced in thin films.

Our purpose in this section is to review some physical effects of possible use in modulators. Particular attention will be paid to electro-optical, acousto-optical, and magneto-optical effects. Although, these are not all suitable for high frequency modulation, they are often useful in measurements and instrumentation. For the sake of completeness, we shall also briefly mention some other, potentially useful, physical phenomena.

We start with electro-optical effects, which are a consequence of the anisotropical nature of certain crystals. In such crystals, the normal relationship between the electric field and its displacement ($D = \varepsilon E$) breaks down, and must be written in tensor form [2.11] as follows:

$$\underline{D} = \underline{\underline{\varepsilon}}\underline{E} \quad ,$$

or,

$$\begin{aligned} D_x &= \varepsilon_{xx}E_x + \varepsilon_{xy}E_y + \varepsilon_{xz}E_z \quad , \\ D_y &= \varepsilon_{yx}E_x + \varepsilon_{yy}E_y + \varepsilon_{yz}E_z \quad , \\ D_z &= \varepsilon_{zx}E_x + \varepsilon_{zy}E_y + \varepsilon_{zz}E_z \quad . \end{aligned} \qquad (2.13)$$

The nine "epsilons" are constants of the medium, and constitute the *dielectric tensor*. Thus, in contrast to the isotropic medium for which $\underline{D}$ and $\underline{E}$ have the same direction, excitation by an electric field in an anisotropic medium produces a displacement which has a different direction. For example, excitation by E_x produces all x, y, z components of D.

With the above definition, the law of conservation of energy can only be satisfied for a symmetric dielectric tensor [2.11]. Under these conditions the connection between $\underline{D}$ and $\underline{E}$ can then be written as

$$\begin{aligned}\underline{E}\cdot\underline{D} &= \varepsilon_{xx}E_x^2 + \varepsilon_{yy}E_y^2 + \varepsilon_{zz}E_z^2\\ &\quad + \left(2\varepsilon_{xy}E_xE_y + 2\varepsilon_{yz}E_yE_z + 2\varepsilon_{zx}E_zE_x\right) = g\ (\text{say}) \quad . \end{aligned} \tag{2.14}$$

The above is the equation of an ellipsoid, which can always be transformed to its principal axes. In other words, there exists a coordinate system in the crystal such that the term in the square brackets disappears, so that only the principal permittivities survive. If we then represent $\underline{E}$ by the position vector $\underline{r}$, and normalize g to unity, we obtain the equation for the *Fresnel ellipsoid*

$$\varepsilon_x x^2 + \varepsilon_y y^2 + \varepsilon_z z^2 = 1 \quad , \tag{2.15}$$

where the second subscript has been dropped for clarity. Since the principal axes of the Fresnel ellipsoid are also the principal axes of the tensor ellipsoid that gives E in terms of D [in contrast to (2.13)], the ellipsoid corresponding to the inverted tensor can also be found. Writing it in terms of the refractive indices gives the so called *index ellipsoid* or *indicatrix* (Fig.2.13):

$$\frac{x^2}{n_x^2} + \frac{y^2}{n_y^2} + \frac{z^2}{n_z^2} = 1 \quad . \tag{2.16}$$

Observe that the semi-axes of this ellipsoid give the principal refractive indices of the crystal. In practice, most of the crystals of interest in electro-optics are *uni-axial*, meaning that they have only one principal axis (z-axis by convention, but also referred to as the c-axis, or simply as the principal axis) along which the permittivity is ε_z, but perpendicular to which ε is independent of direction ($\varepsilon_x = \varepsilon_y$). Such uni-axial crystals have two principal refractive indices ($n_x = n_y$, and n_z). Values of the refractive indices in other directions may be found from the following considerations.

The plane containing the wave vector $\underline{k}$ and the z-axis shown in Fig.2.13 constitutes a plane of symmetry (principal plane) for the ellipsoid. It fol-

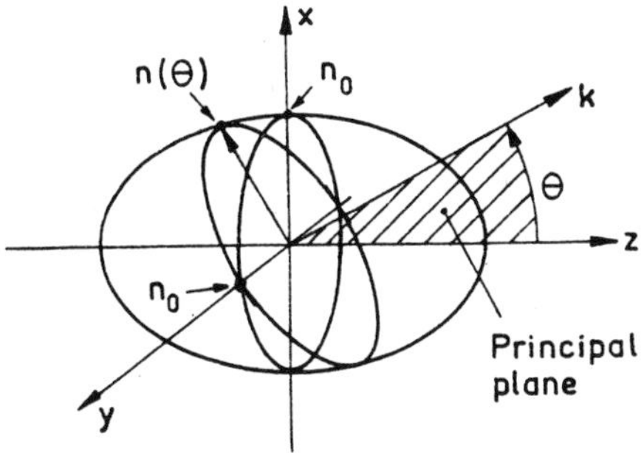

Fig.2.13. The index ellipsoid of an optical crystal

lows that the elliptical section (tilted in Fig.2.13) perpendicular to $\underline{k}$ is also perpendicular to the principal plane, so that the principal axes of the ellipse are perpendicular to and parallel to the principal plane. Thus, if we resolve the field vector of the wave represented by $\underline{k}$ into its two components along the axes of the ellipse, then one of the components is always perpendicular to the principal plane, and the other is parallel to it. In other words, we can resolve the wave into two parts: one vibrating perpendicularly to the principal plane (ordinary wave) and the other in parallel to it (extra-ordinary wave). For any angular direction θ of the wave vector $\underline{k}$, the ordinary wave vibrates along the principal axis, corresponding to $n_x = n_y = n_0$, and therefore experiences the ordinary index of refraction, n_0. The other axis of the ellipse varies with the angle θ, but always remains between the values n_0 and n_z. The latter is usually referred to as the extra-ordinary index of refraction n_e. Hence, for propagation parallel to the z-axis of the crystal, both waves will travel with identical velocities and *birefringence* will be absent. On the other hand, for propagation in a direction perpendicular to the z-axis, birefringence will be at a maximum, and the wave vibrations will be confined to the y (or x) and z-directions. Birefringence is said to be negative for $n_e > n_0$ and positive for $n_e < n_0$, as shown in Table 2.1.

From our viewpoint, the factor of prime interest is that the application of an electrical or mechanical field can cause changes in the index ellipsoid, and can thus enhance or even induce birefringence. Such changes are the immediate result of lattice deformations, which are produced by piezo-electricity and electrical polarization in the one case, and by direct mechanical force in the other. In general, the stress and strain tensors do not have the same principal axes and, in our coordinate system of (2.15) or (2.16), the index ellipsoid assumes a more general form

$$\frac{x^2}{n_1^2} + \frac{y^2}{n_2^2} + \frac{z^2}{n_3^2} + \frac{yz}{n_4^2} + \frac{xz}{n_5^2} + \frac{xy}{n_6^2} = 1 \quad , \tag{2.17}$$

Table 2.1. Birefringence in some uni-axial crystals

Material	Wavelength [μm]	n_e	n_0	$\Delta n = n_e - n_0$
ADP ($NH_4H_2PO_4$)	0.5	1.483	1.530	-0.047
KDP (KH_2PO_4)	0.5	1.472	1.514	-0.042
Quartz (SiO_2)	0.59	1.553	1.544	+0.009
$LiTaO_3$	0.5	2.221	2.216	+0.005
$LiNbO_3$	0.5	2.245	2.344	-0.099
$BaTiO_3$	0.63	2.365	2.416	-0.072
KTN($KTa_{0.65}Nb_{0.35}O_3$	0.55	2.312	2.300	+0.012

which is similar to that in (2.14). Here we have made a replacement of subscripts: xx = 1, yy = 2, etc. The changes produced by an electrical field can be represented in terms of the changes produced in the six terms $(1/n_i^2)$ as follows

$$\Delta(1/n^2)_i = r_{ij}E_j \qquad i = 1, 2, \ldots 6, \quad j = 1, 2, 3, \tag{2.18}$$

where E_j is the applied electric field, and r_{ij}, known as the electro-optic matrix, contains eighteen independent coefficients. For most materials of interest, symmetry considerations reduce this number consideraby [2.12], as can be seen in Table 2.2, which shows the non-zero coefficients.

Table 2.2. Non-zero coefficients in the electro-optic matrix of some commonly used crystals

Material	r_{ij} [10^{-12} m/V]
KDP	$r_{41} = r_{52} = 8.6,\ r_{63} = 9.5$
Quartz	$r_{11} = -r_{21} = -r_{62} = -0.47,\ r_{41} = -r_{52} = 0.2$
$BaTiO_3$	$r_{33} = 28,\ r_{13} = r_{23} = 8,\ r_{42} = r_{51} = 820$
KTN	$r_{33} = 1400,\ r_{13} = r_{23} = 1000,\ r_{42} = r_{51} = 10\ 000$
$LiNbO_3$	$r_{13} = r_{23} = 9,\ r_{22} = -r_{12} = -r_{61} = 6.6,\ r_{42} = r_{51} = r_{33} = 30$

It can be seen from (2.18) that we are considering an effect that is directly proportional to the electric field, and which is known as the *Pöckels effect*. It is obvious that, in the general case, the refractive index change is rather hard to calculate, and that some simplification is required. From a practical viewpoint, it is reasonable to assume that the direction of light propagation is perpendicular to the principal axis. The refractive index change then takes the form

$$\Delta n = -n'^3 r' \frac{E}{2} \quad , \tag{2.19}$$

where n' is a linear combination of the principal refractive indices, r' is a linear sum of electro-optic coefficients, and E is the appropriate component of the applied field [2.10].

Having calculated Δn, the phase shift of the light, after it has travelled a distance ℓ, can be determined from

$$\Delta\phi = \frac{2\pi\ell}{\lambda} \cdot \Delta n \quad . \tag{2.20}$$

Combining (2.19) and (2.20), we can estimate the voltage that must be applied across a width ℓ to obtain a phase shift of π. This voltage, known as the half-wave or $\lambda/2$ voltage, is given by

$$V_{\lambda/2} = \lambda/(n'^3 r') \quad . \tag{2.21}$$

The half-wave voltage varies between about 0.2 and 20 kV for various electro-optical materials. Another figure of merit, the specific energy, is related to the high-frequency performance of electro-optical modulators, and is defined to be the drive power per unit bandwidth, and is obtained by observing that the electrical energy in the electro-optical modulator is given by the well-known relationship

$$W = \frac{1}{2} \cdot \frac{1}{2} \varepsilon E_m^2 \cdot ab\ell \quad , \tag{2.22}$$

where the middle term represents the energy density for peak field strength E_m, and $ab\ell$ gives the volume of the crystal. The factor ½ yields the mean square energy of the temporally harmonic field. Moreover, assuming a one-pole equivalent circuit for the modulator, straightforward circuit analysis shows that the power P in a bandwidth Δf is given by

$$P = 2\pi\Delta f \cdot W \quad . \tag{2.23}$$

Then, use of (2.19), (2.20), and (2.22) in (2.23) yields

$$P/\Delta f = \frac{1}{2\pi} \frac{\varepsilon}{n'^6 r'^2} \lambda^2 (\Delta\phi_m)^2 \frac{ab}{\ell} \quad , \tag{2.24}$$

where $\Delta\phi_m$ is the phase shift corresponding to E_m. For the diffraction-limited case [2.12], we can assume $a = b = \sqrt{4\lambda\ell/(n'\pi)}$, so that substitution into (2.24) gives the minimum specific energy to be

$$P/\Delta f = \frac{2}{\pi^2} \cdot \frac{\varepsilon}{n'^7 r'^2} \lambda^3 (\Delta\phi_m)^2 \quad . \tag{2.25}$$

According to convention, we may set $\Delta\phi_m$ = 2 radians, which corresponds to a modulation depth of 84% [2.10]. Then, for example, the specific energy of $LiNbO_3$ becomes 23 mW/MHz at 630 nm, if we multiply the diffraction-limited dimensions a and b by a safety factor of about 12.

In passing, we note that the phase modulation considered so far can be converted to intensity modulation by using an arrangement such as the one in Fig.2.14a. The active crystal (3) is placed between crossed polarizers (1) and (2), so that the transverse field direction is at 45° to the orientation of the polarizers. The output light is often elliptically polarized even without any external voltage (due to natural birefringence), so that a quarter-wave plate (4) is inserted. Polarizer (2) is then adjusted to achieve a zero output. Upon application of voltage to the electrodes (5), induced birefringence produces some non-zero intensity at the output.

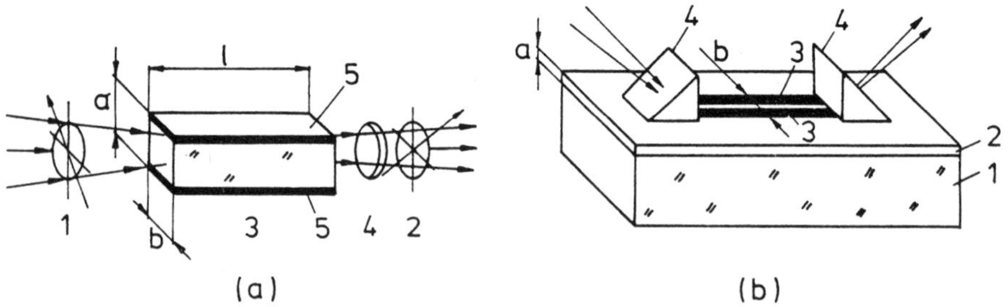

Fig.2.14. (a) Method for converting phase modulation into intensity modulation. (b) Illustration of a channel modulator

The major disadvantage of bulk electro-optical modulators is their relatively high specific energy. As mentioned earlier, this problem can be solved by confining the light within a thin film, as discussed in Sects.3.3 and 3.4. For example, in a strip or channel modulator as shown in Fig.2.14b, the light beam is coupled into a film which is typically 1-50 μm wide and a few micrometers thick. The layer is grown on a substrate (1), the modulating voltage is applied to the pair of electrodes (3), while coupling and decoupling are achieved using special elements, such as the prisms (4). When a voltage is applied, the mode content (see Sect.3.3) of the film is changed due to changes in the refractive index difference between the film and the substrate. Thus, an active substrate instead of an active film could also be used. This results in the phase retardation and/or mode conversion (TE to TM or vice versa) of the propagating light. In any case, the phase and/or amplitude of the output from the film changes according to the applied voltage. It can be shown [2.10], that the specific energy of such a device is a factor of (λ/ℓ) below that given by (2.25). Hence, at a wavelength of 1 μm and $\ell = 0.1$ cm, the specific energy can theoretically be reduced by three orders of magnitude. In practice, the reported figures are 0.76 mW/(MHz) for a non-optimal case [2.13] ($a = 500$ μm, $b = 50$ μm, $\ell = 0.62$ cm), and much better, 5 μW/(MHz), for $b = 5$ μm, and $\ell = 0.3$ cm [2.14]. These results confirm the significant advantage of thin-film modulators, but we must bear in mind that problems of efficient coupling, film imperfections, and general reliability, still need to be solved.

Let us next consider the acousto-optical effect which is, in fact, closely related to the electro-optical effect. The only difference is that changes in the index ellipsoid are caused directly by the mechanical strains due to sound pressure (photo-elastic effect). The refractive index change due to this strain s is given by [2.12,15]

$$\Delta n = -\frac{n^3}{2} ps \quad , \tag{2.26}$$

where p is the strain-optic component of the tensor, similar to $\{r_{ij}\}$. The strain is related to the acoustic power P_a through the acoustic impedance ρv_a^3 as follows [2.12]

$$s = \sqrt{\frac{2\, P_a}{\rho v_a^3 \cdot A}} \quad , \tag{2.27}$$

where A is the area of the transducer, ρ is its density, and v_a is the velocity of the sound in the medium. The index modulation can be viewed as the source term in the wave equation, whose solution (see, e.g. [2.12]) for strong scattering (Bragg regime) can be expressed as in (2.19). Use of (2.26) and (2.27) then gives

$$|\Delta n| = \sqrt{\frac{n^6 p^2}{\rho v_a^3} \cdot \frac{P_a}{2A}} = \sqrt{\frac{M_2 P_a}{2A}} \quad , \qquad (2.28)$$

where M_2 is a figure of merit for acousto-optic materials and is frequently quoted relative of the M_2 value of fused quartz, which has a value of $1.51 \times 10^{-15}\ s^3\ kg^{-1}$ in the longitudinal direction [2.16].

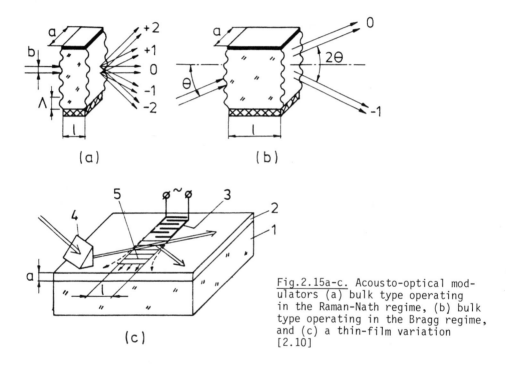

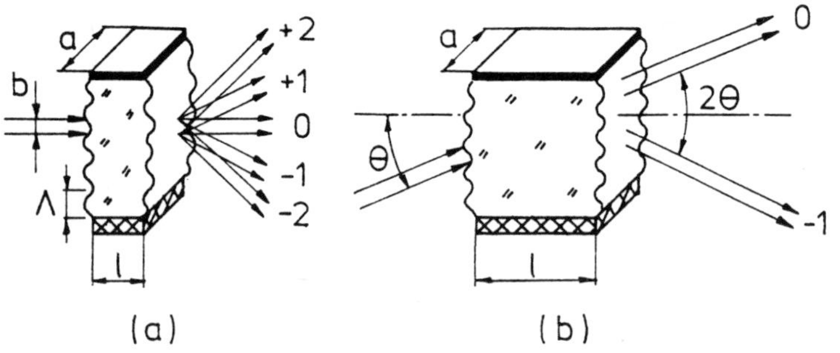

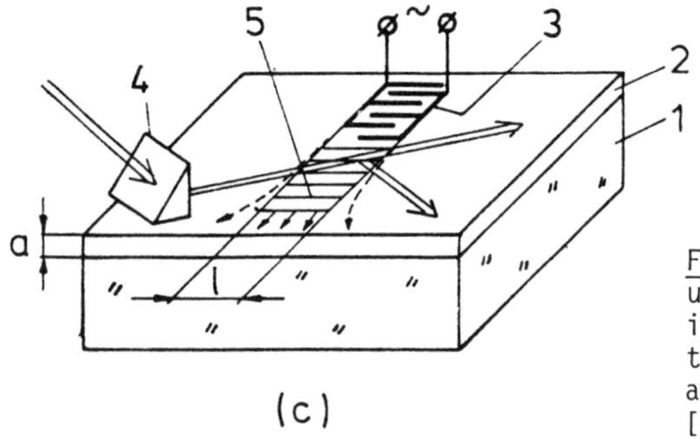

Fig.2.15a-c. Acousto-optical modulators (a) bulk type operating in the Raman-Nath regime, (b) bulk type operating in the Bragg regime, and (c) a thin-film variation [2.10]

From the foregoing, we see that although the physical backgrounds of the electro and acousto-optical effects are similar, modulators based on the latter effect use the feature that the propagating sound wave pattern generates a diffraction grating, as illustrated in Fig.2.15a and b. Two different diffraction regimes can be identified, depending on the modulator thickness ℓ and the acoustical and optical wavelengths Λ and λ, respectively,

1) when $\ell << \Lambda^2/\lambda$, the Raman-Nath diffraction process occurs, and yields a large number of diffraction orders, as shown in Fig.2.15a. The angle and intensity of the mth order is then given by

$$\left.\begin{aligned} \sin\theta_m &= m\lambda/\Lambda \quad , \\ I_m/I_0 &= [J_m^2(\Delta\phi)]/2 \quad \text{for} \quad |m| > 0 \quad , \\ &= J_0^2(\Delta\phi) \quad \text{for} \quad m = 0 \quad ; \end{aligned}\right\} \quad (2.29)$$

where J_m is the Bessel function of order m.

2) When $\ell >> \Lambda^2/\lambda$, we are in the Bragg diffraction regime, and only one diffraction order (besides the zeroth) exists, as shown in Fig.2.15b. Moreover, for $\theta = \theta_B = \arcsin[\lambda/(2\Lambda)]$, this diffraction is very powerful, as can be seen from the intensity values below (normalized to the input intensity)

$$\left.\begin{aligned} I'_m &= \sin^2(\Delta\phi/2) \quad \text{for } m = 1 \quad , \\ I'_0 &= 1 - \sin^2(\Delta\phi/2) \quad \text{for } m = 0 \quad . \end{aligned}\right\} \quad (2.30)$$

We recognize that the power in the input beam can be totally "pumped" into the diffracted beam, in contrast to about 33% for Raman-Nath diffraction. Of course, the frequency of the diffracted beam is shifted from the zero-order frequency by multiples of f_0 (the sound frequency), both for Raman-Nath and for Bragg diffraction. This phenomenon could be used for frequency modulation, but the unshifted, intensity-modulated, zero order is used more often.

The frequency response of Bragg-type modulators is partly affected by the fact that the Bragg angle (hence the diffraction efficiency) depends on the wavelength Λ. However, even more serious is the inherent feature that acoustic waves are slow and require a time of about $\tau = b/v_a$ for penetrating the waist of the light beam. Thus, for rapidly changing modulation signals, the diffraction process becomes "blurred". The specific acoustic energy may then be derived by setting $\Delta f = v_a/b$, and observing that (2.20) and (2.28) yield an acoustic power

$$P_a = \frac{(\Delta\phi)^2\lambda a}{2\pi^2 M_2 \ell} \quad . \qquad (2.31)$$

The specific energy then becomes

$$P_a/\Delta f = \frac{(\Delta\phi)^2\lambda^2 ab}{2\pi^2 v_a M_2 \ell} \quad . \tag{2.32}$$

For diffraction-limited beams, optimum performance is obtained when the optical and acoustical beams have similar waists. Thus, we can set [2.17]

$$\frac{\ell}{b} = \frac{\Lambda n}{\lambda} = \frac{v_a}{f_a}\cdot\frac{n}{\lambda} \quad , \tag{2.33}$$

and subsitution into (2.32) gives

$$P_a/\Delta f = \frac{(\Delta\phi)^2 a\lambda^3 f_a}{2\pi^2 n v_a^2 M_2} \quad . \tag{2.34}$$

By further requiring the transducer dimension a to be optimum, i.e. $a = 2\Lambda/\sqrt{\pi} = 2v_a/(\sqrt{\pi}f_a)$, we obtain

$$P_a/\Delta f = \frac{(\Delta\phi)^2}{\pi^{2.5}}\cdot\frac{\lambda^3}{n v_a M_2} \quad . \tag{2.35}$$

Typical values of the specific energy for $\Delta\phi = 2$ radians are shown in Table 2.3. We observe that the specific energies of bulk acousto-optical modulators is significantly lower than those of the electro-optical counterparts. However, this advantage is often counterbalanced by the practical problem of being able to transduce strong acoustical waves from electrical signals.

Table 2.3. Selected properties of some acousto-optical materials

Material	Index, n (λ = 633 nm)	Acoustic velocity [m s^{-1}]	M (rel. to SiO_2)	Δn ($P/A = 10^6$ Wm^{-2})	$P/\Delta f$ [mW/MHz]
TeO_2	2.27	0.617×10^3	525	6.3×10^{-4}	0.05
$PbMoO_4$	2.39	3.66×10^3	23.7	1.34×10^{-4}	0.18
$LiNbO_3$	2.2	6.57×10^3	4.6	0.58×10^{-4}	0.59
SiO_2 (fused)	1.46	5.59×10^3	1.0	0.27×10^{-4}	4.41

We now return to the question of the bandwidth of acousto-optical modulators, and observe that, as in the electro-optical case, the bandwidth can be improved by the use of thin films. Figure 2.15c shows the basic arrangement of such modulators. The light is launched by the prism (4) into the surface waveguide (2), which is deposited on the substrate (1). Surface acoustical waves are generated by the transducer (3). The important feature here is that the surface layer, which has a thickness of about Λ, confines both the acoustical and optical waves, thereby increasing the modulation efficiency. The specific energy in this case [2.18] can be estimated from (2.34) by dividing its right hand side by a factor ξ. This term (called an overlap factor) accounts for the fact that a non-uniform acoustic wave interacts with a non-uniform optical wave. When the acoustic strain field overlaps congruently the optical field, ξ is equal to one, and is less than one for all other cases [2.10]. In practice, a value of about 0.2 mW/MHz was reported by OHMACHI in 1973 [2.19].

Our next task is to take a brief look at magneto-optical modulators, which can use a variety of effects for modulating some optical parameter such as the refractive index [2.20]. Of these, the most important is the Faraday effect, which leads to specific energies comparable to those of electro-optical and acousto-optical transducers. In its simplest form, the Faraday effect can be described as a rotation of the polarization direction of linearly polarized light upon application of a magnetic field. In general, the field is used to change the direction or magnitude of the bias magnetization or alternatively the internal magnetization already existing in the magneto-optic material. Practically, a bulk modulator such as that in Fig.2.16a consists of a cylinder of magneto-optic material (e.g., doped YAG) surrounded by a coil. When an electrical current is passed through the coil, a longitudinal magnetic field is produced. If linearly polarized light is passed through

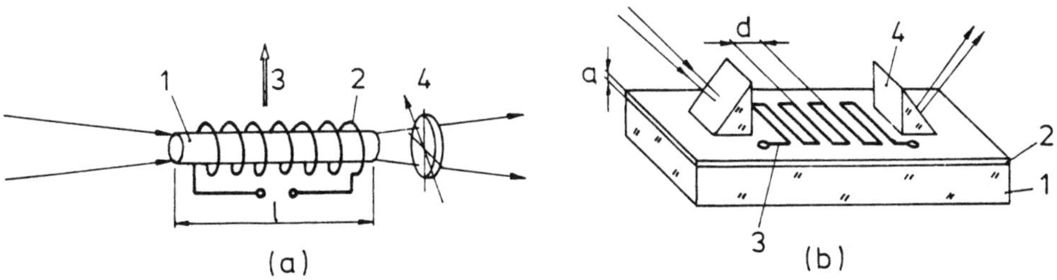

Fig.2.16. (a) Depiction of a bulk magneto-optical modulator. (b) A thin-film magneto-optical modulator

the cylinder, the polarization direction changes in sympathy with the magnetization (electrical current), so that the output optical power (after the analyzer) also follows the electrical current.

In a gyromagnetic medium, with a saturated transverse field, the magnetization produced by the longitudinal modulating field is given by [2.21]

$$M_x = \chi \cdot H_x \approx \frac{H_x}{H_i} M_s \quad , \tag{2.36}$$

where H_i is the initial magnetizing field, H_x is the peak longitudinal field, χ is the susceptibility, and M_s is the saturation magnetization. Moreover, the Faraday rotation angle is characterized by the equation [2.21]

$$\theta = \frac{M_x}{M_s} \cdot \theta_F \cdot \ell \quad , \tag{2.37}$$

where θ_F is the rotation per unit length for saturation magnetization, and ℓ is the length of the sample. Using (2.36) and (2.37), we can now evaluate the magnetic energy stored in the coil, from the formula [cf.(2.22)]

$$W = \frac{1}{2} \cdot \frac{\pi d^2 \ell}{4} \cdot \frac{1}{2} H_x B_x \quad , \tag{2.38}$$

where $B_x = \mu_0(H_x + M_x) = \mu_0 H_x(1+\chi)$, $\mu_0 = 4\pi \times 10^{-7}\ \mathrm{H\,m^{-1}}$ is the permeability of free space, and the middle term represents the volume of the cylindrical rod. Then, using (2.23), the specific energy is found to be

$$P/\Delta f = \frac{\pi^2 d^2}{8\mu_0} \cdot \frac{(1+\chi)}{\chi^2} \left(\frac{\mu_0 M_s}{\theta_F}\right)^2 \frac{\theta^2}{\ell} \quad . \tag{2.39}$$

If we now set $\theta = 2$ radians, and assume a diffraction-limited beam [2.12] with $d^2 = 4\lambda\ell/(\pi n)$, we obtain

$$P/\Delta f = \frac{2\pi\lambda}{n\mu_0} \cdot \frac{(1+\chi)}{\chi^2} \left(\frac{\mu_0 M_s}{\theta_F}\right)^2 \quad . \tag{2.40}$$

In (2.39) and (2.40), $\mu_0 M_s$ has the dimensions of Tesla ($\equiv 10^4$ Gauss), while θ_F has the dimensions of radians/meter. It can be seen that the minimum specific energy is directly proprotional to λ, as opposed to λ^3 in (2.25) and (2.35).

The thin-film variation of the magneto-optical modulator is shown in Fig. 2.16b. In this case, the magnetization orientation is at some angle ($\sim 45°$)

to the direction of light propagation. The magnetic field is generated by the serpentine structure of period d. The period is chosen such that the corresponding angular frequency, Δ, is given by

$$\Delta = 2\pi/d = \beta_{TE} - \beta_{TM} \quad , \tag{2.41}$$

where β_{TE} and β_{TM} are the propagation constants of two neighbouring TE and TM modes (Sect.3.3). As a result, the vector of the magnetic field in the direction of propagation contains a component with a spatial periodicity d, and strong mode conversion between TE and TM modes occurs. Because of the high birefringence of rutile, TE and TM modes emerge at very different angles (e.g., TE_0 and TM_0 modes have a separation around 20°). Thus, any modulation of the mangetization current can be observed as a power change in one of the modes.

The specific energy for couplers of this type can be estimated from (2.39) by replacing the area $\pi d^2/4$ by ab and assuming that a is independent of ℓ, while b is diffraction limited according to $b = \sqrt{2\ell\lambda/n}$ [2.10]. The following result is then obtained

$$P/\Delta f = \frac{2\pi a}{\mu_0}\sqrt{\frac{2\lambda}{n\ell}}\,\frac{(1+\chi)}{\chi^2}\left(\frac{\mu_0 M_s}{\theta_F}\right)^2 \quad . \tag{2.42}$$

In practice, experimental values in the order of 1 mW/MHz have been achieved [2.22], and do not represent a significant improvement over the specific energies of bulk modulators.

The most serious problem associated with magneto-optical modulators is that the absorption in most of the usable materials is rather high in the visible and near infrared regions. However, this problem is less serious at, say, 1.3 or 1.55 μm, which are important transmission wavelengths in fiber systems. Hence, magneto-optical modulators may provide a viable alternative in some cases.

We conclude this section by listing a few other effects that could conceivably be used for the modulation of light.

1) In general, the response of the index ellipsoid to an external electric field is far more complex than that represented by Pöckel's formula (2.19). To some extent, we can account for this complexity by writing Pöckel's formula in a series form:

$$\left.\begin{aligned} n(E) &= n(E_0) + (E - E_0)\frac{dn}{dE} + \frac{(E - E_0)^2}{2}\frac{d^2n}{dE^2} + \dots , \\ &\text{or} \\ n &= a_0(\Delta E) + a_1(\Delta E)^2 + \dots , \end{aligned}\right\} \qquad (2.43)$$

where E is the applied field, E_0 is the internal field of the crystal (which, for example, leads to natural birefringence), while a_0, a_1 ... are constants. In certain types of centro-symmetric crystals, such as $BaTiO_3$ and KTN, the index change can be adequately described by just the second-order term. This type of behaviour is known as the Kerr effect, and can be useful for "on-off" modulation.

2) In certain types of ceramics (mostly La-doped $PbZr_xTi_{1-x}O_3$, so called "PLZT" ceramics) electro-optical anisotropy occurs due to the existence of ferro-electric domains. In crystals with well-defined domains, an external field can be used to change the index ellipsoid by tilting the polarization of the domains. As a result, the electro-optical behaviour in strongly polarized ceramics resembles the magneto-optical effect.

3) In a second class of ceramics, the observed change in refractive index (upon application of an electric field) resembles the Kerr effect even though the physical reason is different [2.23].

4) A third class of ceramics exhibits the effect of "dynamic scattering", which is rare in solids, being more associated with liquid crystals. Under the influence of an external voltage, a significant change in the size and orientation of the ferro-electric domains takes place. Due to the large refractive index and strong birefringence of individual domains, the process of optical scattering at the domain boundaries becomes an important mechanism, particularly when the domain cross-section increases to the order of magnitude of the wavelength.

5) Except for dynamic scattering, all the effects we have discussed so far are used for modulating the phase. Obviously, it would be useful to have a modulator that could directly affect the intensity, and thus eliminate the use of polarizers, etc. One possible candidate may be the Franz-Keldysh effect [2.24,25], which has been observed in the important semiconductors used in infrared devices (Ge, Si, GaAs). This is the effect whereby, upon application of a sufficiently high electric field, a shift in the absorption edge takes place (Sect.2.3). Thus, an external field could be used for changing the absorption and hence the intensity.

6) In some materials (mostly ferro-electrics), due to the abrupt electric field change at the surface, the surface or Känzig layers [2.26] have a refractive index which is different from that of the bulk material. (This abrupt change occurs at distances defined by the screening parameter, which reflects the ability of a solid to re-arrange its distribution of space charge). In any case, the thickness of the Känzig layers and the refractive index difference from the bulk material allow us to hope in the possibility of "pre-fabricated" low-mode waveguides, whose mode content could be modulated by an external electric field.

2.3 Detection of Light

The two physical phenomena that are most commonly used for the detection of light are [2.27]: 1) the temperature increase of a small mass due to incident radiation, and 2) the photo-electric effect, whereby a quantum of light of sufficiently high energy causes the release of an electron in the material. The photo-electric effect can actually be used in two ways: I) The incident radiation falls on a material (in a vacuum) from which surface electrons are released for collection (vacuum photo-diode), and II) the incident radiation creates free charge carriers (in the material), which lead to a current flow through the material. From the viewpoint of fiber-optical systems, this latter represents the most widely used phenomenon, because of the small size and high speed that can be achieved using this approach. In this section, we will restrict ourselves to a discussion of the physical principles behind the operation of such detectors [2.28].

Let us consider what happens when a semiconductor is externally irradiated by a source of variable frequency ν, or photon energy $h\nu$. Neglecting any non-linear optical processes, we see that no transition can occur as long as $h\nu < \Delta E$, and the material appears transparent to the radiation (no absorption). Of course, this does not mean that the radiation does not "sense" the material. The various charges that are contained in the solid will *all* respond, in some measure, to the incident electromagnetic wave. The electric field of the wave will force displacement of the charges with respect to their initial positions, and thus produce dipoles. The net polarization produced will be a sum total of the unequal individual contributions of all charges. The tightly bound electrons making up the core of an ion will only be weakly polarized, whereas weakly bound outer-shell electrons will contribute much more. Furthermore, since the electric field of the radiation is inevitably oscillatory, the induced dipoles will also try to follow the fluctuations. The heavy ions are only able to follow at or below infrared frequencies, while electrons start to experience difficulty at frequencies corresponding to the ultraviolet. If it is assumed that the energy lost by the oscillating dipoles is negligible, the electrons will re-radiate at the same frequency as that of their excitation. In other words, the presence of charges and their interaction with the light wave will not cause any energy loss from the wave, but will merely delay its transmission. This will naturally reduce the effective velocity of the light. Hence, the refractive index of the material will be greater than unity since

$$n = c/v \quad . \tag{2.44}$$

Here, c is the velocity in vacuum, while v is the velocity in the material. The refractive index is thus a result of the polarizability of the material. The larger the polarization, the more will be the delaying action, and the larger the refractive index. For materials that have more massive atoms and hence more electrons, the polarizability is enhanced, which yields higher refractive indices. As an example, the heavy glasses always have higher refractive indices than the lighter ones (Fig.2.17). The refractive indices of some materials of interest are given in Table 2.4.

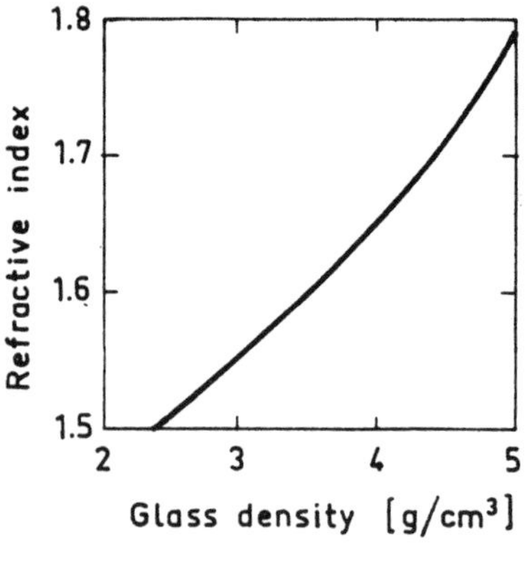

Fig.2.17. Refractive index of glass as a function of density

Let us now increase the frequency of our monochromatic source such that the photon energy $h\nu$ becomes greater than the band gap. In this case, the photon can excite an electron from the valence band to the conduction band (absorption). In fact, every semiconductor becomes absorptive at some critical wavelength. This limit, known as the absorption edge, is

$$\lambda_c = \frac{hc}{\Delta E} \quad , \tag{2.45}$$

and forms an important characteristic from the viewpoint of optical behaviour. The absorption edge values of some common semiconductors are shown in Table 2.5.

Table 2.4. Refractive indices of some materials

Material	Refractive index	Wavelength [μm]
Quartz	1.54	0.59
Glasses	1.5 - 1.8	0.59
ZnS	2.37	0.59
Ge	4.09	2.15
Si	3.45	2.15
GaAs	3.3	5.0
PbS	4.1	3.0

Table 2.5. Absorption edge values of some common semiconductors at 300 K

Material	Energy gap, ΔE [eV]	λ_c [μm]
PbS	0.34 - 0.37	3.35 - 3.65
Ge	0.67	1.85
Si	1.14	1.09
GaAs	1.43	0.87
GaP	2.26	0.55

For semiconductors with more than one valley in their conduction band (see Fig.2.5), we have a number of band-gap minima, so that we have to decide which one is appropriate for the definition of the absorption edge. The problem is resolved if we consider that the momentum of a photon is small compared to that of an electron. The most likely transitions must then correspond to a zero change in momentum, in order to satisfy conservation of momentum requirements. An absorption edge defined in this manner would represent the threshold of intense optical absorption. However, indirect transitions at lower photon energies are not ruled out, because momentum can also be conserved by the creation of a phonon:

$$k\ (\text{photon}) = k_c \pm K \approx 0 \quad , \tag{2.46}$$

where k_c is the momentum separation between the top of the valence band and the indirect gap, while K is the phonon momentum. Phonon energies are characteristically small, so that momentum can be rather economically preserved in this way. The absorption due to direct and indirect transitions is shown schematically in Fig.2.18, from which we see that, to define the absorption edge as the limit at which the material becomes transparent, the absorption edge should be given in terms of the minimum band gap, as in Table 2.5. Such a definition is quite practical and reflects on the limit of usefulness of the material (as a detector).

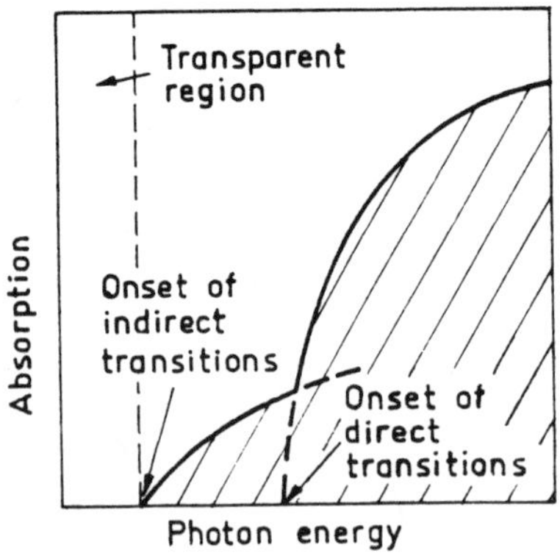

Fig.2.18. Schematic illustration of absorption due to direct and indirect transitions

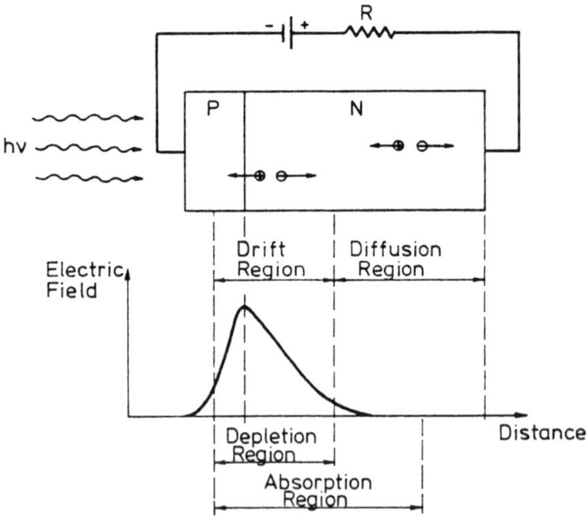

Fig.2.19. Schematic diagram of a p-n junction photodiode

Let us now consider a reverse biased p-n junction (Sect.2.1) of a material that has adequate absorption at the wavelength of interest. Such a junction is shown schematically in Fig.2.19. Photons incident on the p-side are absorbed both in the depletion region and outside of it. Consequently, some electron-hole pairs are generated within the high field region (shown in Fig.2.19) and are accelerated by the field to produce a "drift" current. Other electron-hole pairs are generated in a region of low or zero field and can only cause a "diffusion" current. At a given wavelength, the ratio of these two types of carriers is determined by the width of the depletion region, which, in turn, depends on the reverse bias and the doping levels (the lighter the doping level, the wider the depletion region). Because diffusion is very slow compared to drift, the response time of such p-n junctions depends strongly on the number of diffusion carriers. Normally, such diodes are not of great interest in communication systems.

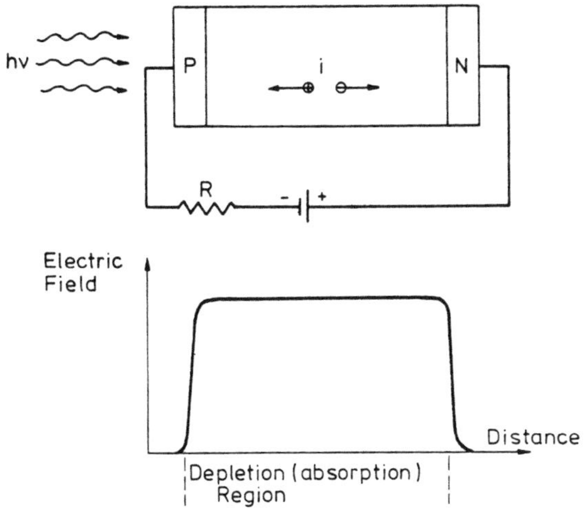

Fig.2.20. Schematic diagram of a p-i-n junction photodiode

The usual means of resolving the above problem is to use a p-i-n structure, shown schematically in Fig.2.20. The "i" or intrinsic region can now be tailored to ensure that most of the absorption takes place within it. Because intrinsic semiconductors have a negligibly small number of free carriers, the intrinsic region experiences a relatively high field, and all the carriers generated within it are accelerated by the field and contribute to the drift current. It follows that p-i-n diodes can operate much faster than "diffusion" diodes.

We see that both in p-n and in p-i-n junctions, irradiation yields a flow of current (diffusion and/or drift) that can be sensed in the external circuit (e.g. as a voltage change across the resistor R in Figs.2.19 and 2.20). The magnitude of the current is determined by the quantum efficiency η (the probability of electron-hole pair generation) of the diode structure. If the incident optical power is P, then the generated current I can be written in the form

$$I = \eta \cdot \frac{qP}{h\nu} \quad , \tag{2.47}$$

where η is the quantum efficiency at optical frequency ν, and q is the electronic charge. In the above, $P/h\nu$ represents the average number of incident photons per unit time, while I/q is the corresponding number of electron-hole pairs. Evidently, to maximize the current, we should maximize the quantum efficiency by selecting a suitable semiconductor, and the optimum structure (e.g. width of "i" region). Typical curves for silicon and germanium are shown in Fig.2.21, together with the emission wavelengths of various lasers. It should be noted that the quantum efficiencies decrease at shorter wavelengths. On the basis of our simple description, this is rather unexpected, and is due to the fact that at shorter wavelengths photons are absorbed very near the surface where the recombination time of electrons and holes becomes shorter than the time required to sweep them out.

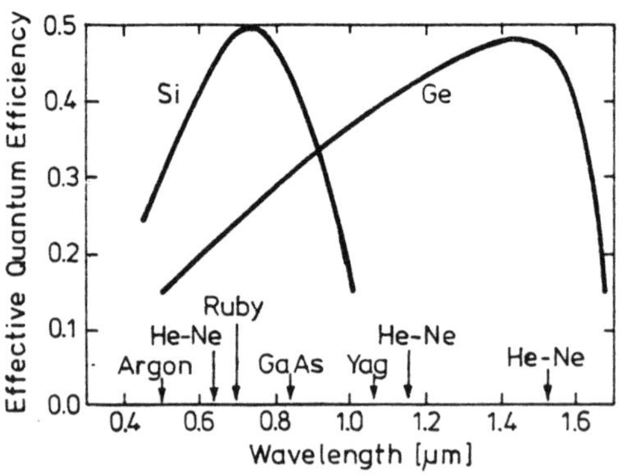

Fig.2.21. Effective quantum efficiency of silicon and germanium photo-detectors

In a p-i-n diode it is obvious that, the higher the drift field (reverse bias), the higher will be the acceleration of the charge carriers, until eventually they will have sufficient energy to release electrons (holes) by impact ionization. Normally, this process leads to avalanche "breakdown" (as in other semiconductor components), but in so-called avalanche photodiodes it can be used as a mechanism for internal current gain (amplification). The schematic diagram of such a structure is shown in Fig.2.22. It can be seen that, as in the p-i-n diode, we now have an intrinsic region where electron-hole pairs are generated, and a high-field region in which avalanche multiplication takes place. The basic requirement for the production of useful current gain in such a device is the elimination of micro-plasmas, i.e. small areas in which the breakdown voltage is less than that of the junction as a whole. This is normally achieved by the use of substrate materials of uniform doping concentration, low dislocation density, and by designing the active area to be no larger than necessary to accomodate the light beam. Moreover, to prevent edge breakdown, some form of guard ring that suitably lowers the field is normally used.

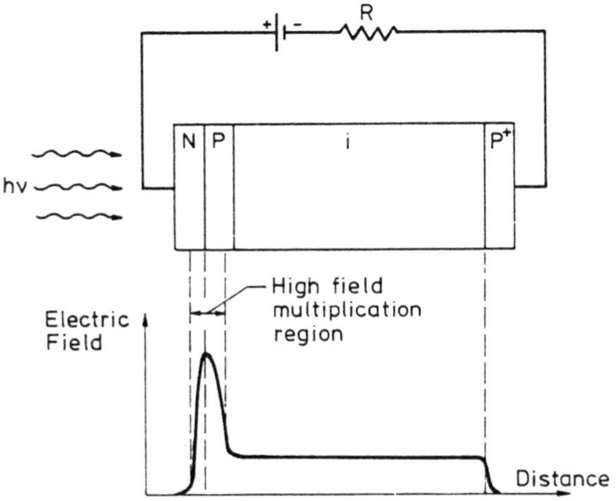

Fig.2.22. Schematic diagram of an avalanche photodiode

The multiplication factor M of avalanche photodiodes is defined to be the ratio of the currents with and without multiplication, and can be described by the formula

$$M = I_m/I_0 = [1 - (V/V_B)^n]^{-1} \quad , \qquad (2.48)$$

where I_m is the multiplied current, I_0 is the current at low voltages (no multiplication), V_B is the breakdown voltage, and n is a constant that depends on the semiconductor material, doping profile, and the radiation wavelength. The current flow after multiplication can be obtained by observing that I_0 in (2.38) must be the same as I in (2.37). Thus, we obtain

$$I_m = \eta \cdot \frac{qP}{h\nu} \cdot M \quad . \tag{2.49}$$

From the viewpoint of circuit behaviour, this equation emphasizes the fact that avalanche and p-i-n diodes are similar, and become identical when the avalanche diode is operated at low voltages (M = 1). An important aspect in which they differ is their noise behaviour. The avalanche process, being statistical in nature, introduces randomness in multiplication. In other words, each primary pair is replaced by a random number of secondary pairs, whose variance is large (for details see Sect.4.5). In designing the photodiode, the randomness is minimized by selecting uniform materials in order to ensure that multiplication is not too random a function of the path taken by the carriers. Moreover, as multiplication statistics are also a function of the location where the primary electron-hole pair is created relative to the high-field region, the device is usually designed such that most primary pairs are created in the intrinsic region (as in p-i-n diodes). Then, only the more ionizing carriers (electrons in Fig.2.22) will drift into the high field for multiplication. Carriers that are undesirably created in the low- or zero-field regions, and are candidates for multiplication, represent a problem because they must first diffuse to the high-field region. In terms of temporal behavior, such carriers produce "diffusion tails" in the impulse response, because the diffusion current continues to flow after the optical radiation has been switched off.

In any case, it is now clear that, while the avalanche photodiode is only a p-i-n diode with a separate multiplication region, its overall behaviour does differ somewhat. It should also be noted that other types of depletion layer photodiodes (e.g. see [2.3]), such as the metal-semiconductor (Schottky barrier), the heterojunction, and the point contact, are also possible. As to usable materials, the best known is silicon, with reasonable quantum efficiencies up to about 1100 nm (Fig.2.21). In fact, given the well-established position of silicon technology, optical fiber communication systems in the range 800 to 1100 nm are almost exclusively based on silicon detectors. As we have already seen, silicon becomes transparent beyond 1100 nm, and one has to resort to other semiconductors. Germanium can be used up to about 1600 nm, but suffers from the disadvantage of poor noise performance at room temperatures.

Cooling alleviates this problem to some extent, but is not really practical for field use. Finally, we should note that the longer-wavelength region is particularly interesting (as will be seen in later chapters) and has motivated the search for ever better detector materials. However, we should bear in mind that the longer wavelengths inevitably require a lower band-gap material, so that the probability of random thermal excitation (noise) is bound to be higher than at shorter wavelengths.

In summary, we have presented a semi-qualitative review of the physical principles behind the generation, modulation, and detection of light. For more detailed treatments of the physics, we refer the reader to the cited works, while for devices particularly intended for fiber-optics, we suggest [2.29].

3. Light Propagation in Waveguides

As is well known, the basic process of light propagation in optical waveguides is due to the total internal reflection of the light at two or more interfaces. In the case of two parallel and abrupt interfaces the process is simply the zig-zag reflection of the light beam, while in the case of a graded refractive-index profile we can understand propagation by considering the grading to occur in a large number of small steps. Then a light beam that enters the waveguide at its center (maximum refractive index) will at first be refracted a number of times, and will eventually be turned around by total internal reflection. Thus, the beam can be expected to follow a smooth and periodic path.

Although the behavior of optical waveguides can be almost completely analyzed on the basis of Maxwell's equations, from the viewpoint of an intuitive physical understanding, it is important to be familiar with the phenomenon of total internal reflection. Towards this end, we devote Sects.3.1 and 2 to a review of the behavior of plane waves and bounded beams at plane, curved, blurred, and periodically modulated interfaces. Since the treatment is intended to be explanatory rather than rigorous, plausibility-type arguments are freely used, while most of the analytical details can be found in the references quoted in the appropriate places.

The above approach is insufficient for the design of communication systems, and we must also consider the effect of propagation delays on the impulse response or bandwidth of the fiber. These questions are treated in Sects.3.5 and 6 via the solution of Maxwell's equations, but to lay the foundations for the analyses, we first consider the behaviour of the analytically simpler planar waveguides (Sects.3.3,4).

Finally, in Sect.3.7, we briefly discuss the problem of speckle in multimode fibers, in preparation for the treatment of analog systems (Chap.6) in which speckle leads to the phenomenon of modal noise.

3.1 Total Internal Reflection at a Plane Interface Between Two Media

We start with the plane interface between two homogeneous, isotropic and lossless media, and observe that the laws of reflection and refraction can be easily obtained by the direct application of Maxwell's equations [3.1]. The usual assumptions are: 1) that the incident radiation is a plane wave, 2) that the problem is independent of one Cartesian coordinate, and 3) that the radiation is polarized either parallel to the plane of incidence (H polarization) or perpendicular to it (E polarization). Under these assumptions, the following results are obtained [3.1]

$$\theta_i = \theta_r \quad , \tag{3.1}$$

$$n \sin\theta_t = \sin\theta_i \quad , \qquad (n = n_2/n_1) \quad , \tag{3.2}$$

$$\left.\begin{aligned} R_\perp &= \frac{E_r}{E_i} = \frac{\cos\theta_i - n\cos\theta_t}{\cos\theta_i + n\cos\theta_t} \quad , \\ T_\perp &= \frac{E_t}{E_i} = \frac{2\cos\theta_i}{\cos\theta_i + n\cos\theta_t} \quad , \end{aligned}\right\} \quad \text{(E polarization)} \tag{3.3}$$

$$\left.\begin{aligned} R_\parallel &= \frac{n\cos\theta_i - \cos\theta_t}{n\cos\theta_i + \cos\theta_t} \quad , \\ T_\parallel &= \frac{2\cos\theta_i}{n\cos\theta_i + \cos\theta_t} \quad , \end{aligned}\right\} \quad \text{(H polarization)} \tag{3.4}$$

where E_r, E_i, and E_t refer to electric fields, while R and T are the coefficients of reflection and transmission (refraction). The other symbols are indicated in Fig.3.1. Equations (3.1.) and (3.2) are the well-known laws of reflection and refraction (or Snell's law), while (3.3) and (3.4) are Fresnel's formulae for E and H polarizations, respectively. Typical sketches of $R_\parallel$ and $R_\perp$ for $n_1 > n_2$ are shown in Fig.3.2., from which we see that, for both polarizations, reflection becomes total at $\theta_i \geqq \theta_{ic}$. At $\theta_i = \theta_{ic}$, (3.2) still defines an angle of refraction of $\theta_t = \pi/2$, but according to (3.3) and (3.4) the wave in this direction has zero power. This is not strictly true, and it will be seen later that an evanescent wave occurs in this direction.

From the foregoing, it is apparent that an interface between two media with different refractive indices can be used as a perfect reflector, at least in the case of a plane wave. However, in practice we must necessarily use

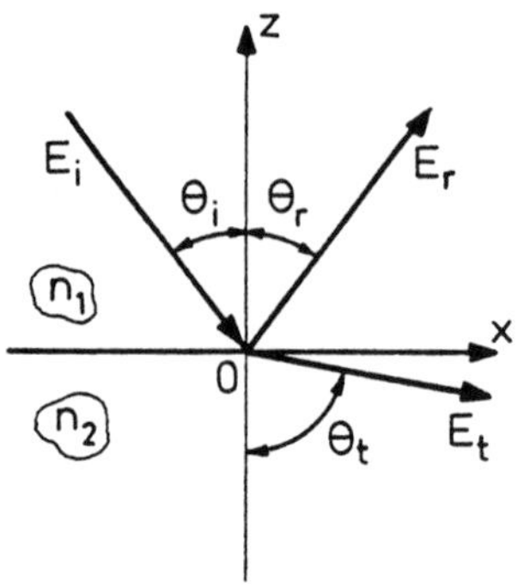

Fig. 3.1. Reflection and refraction at an interface between a dense and less-dense medium

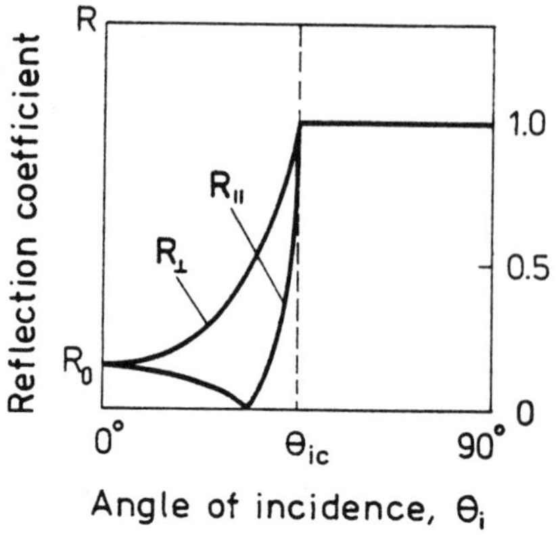

Fig. 3.2. Dependence of reflection coefficients $R_{\parallel}$ and $R_{\perp}$ upon the angle of incidence

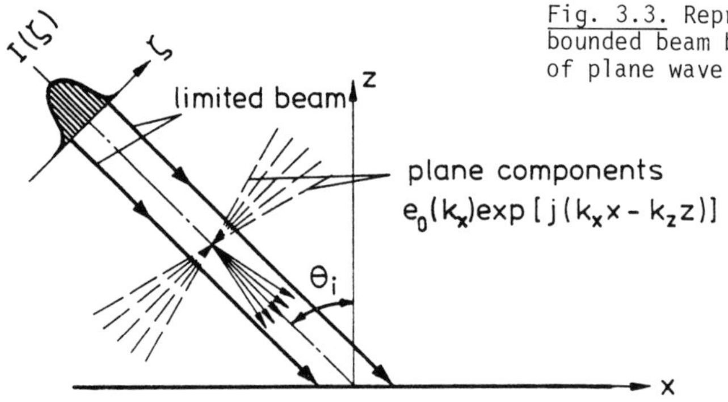

Fig. 3.3. Representation of a bounded beam by an equivalent set of plane wave components

limited beams rather than infinitely spread plane waves, and we would like to know whether perfect reflection is also possible in the case of limited beams. One way to exploit the easily understandable properties of plane waves for beams, is to consider the beam as a superposition of a set of plane waves [3.2-4], in much the same way as electrical signals are transformed into sine-wave Fourier components. In other words, a beam such as the one depicted in Fig.3.3, can be represented as a composition of plane waves, each with wave vector k_x (or angle of incidence θ), and amplitude $e_0(k_x)$, such that

$$E_i(x,z) = \frac{1}{2\pi} \int_{-\infty}^{\infty} e_0(k_x) \exp[j(k_x x - k_z z)] dk_x \quad , \tag{3.5}$$

where a time dependence $\exp(-j\omega t)$ has been assumed. Here, k_x and k_z are the propagation vectors in the x and z directions, respectively, and are related to the vector (k_1) in the direction of propagation by the usual separation condition for vectors:

$$k_z^2 = k_1^2 - k_x^2 \quad , \tag{3.6}$$

where $k_1 = 2\pi n_1/\lambda$.

Consider, for example, a symmetrical beam, as in Fig.3.3, whose central component is incident critically, i.e. at $\theta_i = \theta_{ic}$. Then, components with $\theta_i > \theta_{ic}$ will be perfectly reflected, while components with $\theta_i < \theta_{ic}$ will be reflected with a rapidly decreasing coefficient, as indicated in Fig.3.4. Thus, in contrast to the geometrical optics picture, Fourier optics predicts possible loss from a beam that is incident at the critical angle. Consequently, the plane-wave composition of the reflected beam will be distorted, and can be written in the form

$$E_r(x,z) = \frac{1}{2\pi} \int_{-\infty}^{\infty} R(k_x) e_0(k_x) \exp[j(k_x x + k_z z)] dk_x \quad , \tag{3.7}$$

where $R(k_x)$ is the reflection coefficient of each plane wave component. On the basis of these arguments, it is clear that the "weight center" of the reflected beam must be shifted to the right of the point of incidence. This phenomenon is known as the Goos-Hänchen shift, in recognition of the scientists who demonstrated it experimentally in 1947 [3.5].

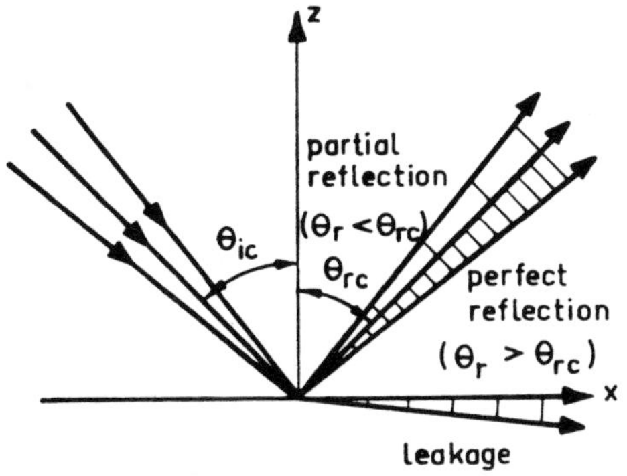

Fig. 3.4. Energy leakage through the interface, for a ray complex that represents plane wave components of a bounded beam

Any precise description of the Goos-Hänchen shift would require the solution of Maxwell's equations, which is difficult for the general case. However, the difficulty can be somewhat eased by choosing a suitable beam profile. A formulation that has been used by LOTSCH [3.6-8], assumes a beam whose transverse amplitude function $A(\beta)$ varies slowly and satisfies the condition

$$A(\beta) \gtrsim \frac{dA}{d\beta} \gtrsim \frac{d^2A}{d\beta^2} \quad , \tag{3.8}$$

where $\gtrsim$ means "at most of the order of magnitude of", and β is given by

$$\beta = \frac{2\pi}{\lambda_0} \zeta\alpha \quad . \tag{3.9}$$

Here, α is a small constant ($\ll 1$), and ζ is the transverse coordinate (Fig. 3.5), related to the Cartesian coordinates by

$$\zeta = x \cos\theta + z \cos\theta \quad . \tag{3.10}$$

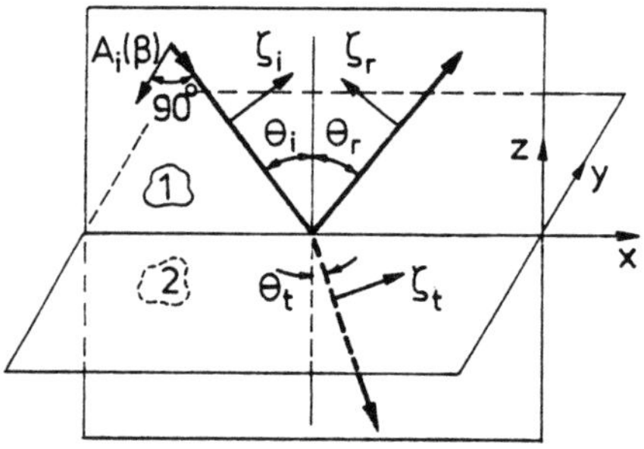

Fig. 3.5. The system of local coordinates used for representing the interaction between a bounded beam and a plane interface [3.6]

If we assume that the solutions of Maxwell's equations for the incident, reflected, and transmitted beams can be approximated by

$$\left.\begin{aligned} E_i &= A(\beta_i)\, e^{j\phi i} - j\alpha_1 a_\perp \left(\frac{dA}{d\beta_i}\right) e^{j\phi i} \quad , \\ E_r &= R_\perp A(\beta_r)\, e^{j\phi r} - j\alpha_1' r_\perp \left(\frac{dA}{d\beta_r}\right) e^{j\phi r} \quad , \\ E_t &= T_\perp A(\beta_t)\, e^{j\phi t_\perp} - j\alpha_2 t_\perp \left(\frac{dA}{d\beta_t}\right) e^{j\phi t} \quad , \end{aligned}\right\} \tag{3.11}$$

then the boundary conditions (continuity of tangential field components across the interface) require that

$$a_{\perp} = -\tan\theta_t \cdot \frac{\cos\theta_i - n\cos\theta_t}{\cos\theta_i + n\cos\theta_t} \quad , \tag{3.12}$$

$$r_{\perp} = \tan\theta_t \cdot \frac{3\cos\theta_i + n\cos\theta_t}{\cos\theta_i + n\cos\theta_t} R_{\perp} \quad , \tag{3.13}$$

$$t_{\perp} = -\frac{\tan\theta_i}{\cos\theta_i} \cdot \frac{2\cos\theta_i + n\cos\theta_t}{\cos\theta_i + n\cos\theta_t}(\cos\theta_i - n\cos\theta_t)T_{\perp} \quad . \tag{3.14}$$

In the above, the beams are assumed to be E polarized and to have incident, reflected, and transmitted peak amplitudes of unity, $R_{\perp}$, and $T_{\perp}$, respectively. The terms ϕ_i, ϕ_r, and ϕ_t represent the corresponding oscillating phases, while α_1, α_1', and α_2 are given by

$$\alpha_1' = -\alpha_1, \quad \alpha_2 = \alpha_1 \cos\theta_i(n^2 - \sin^2\theta_i)^{\frac{1}{2}} \quad , \tag{3.15}$$

where α_1 is the small constant of (3.9), for $\beta = \beta_i$.

Without going into further detail, for which we refer the reader to [3.8], we can now make a number of interesting observations. First, we note that the forms in (3.11) have been chosen such that for a constant $A(\beta)$, i.e. $dA/d\beta = 0$, the second term in each equation disappears, and the field components reduce to the normal form for plane waves. In this case, the ratios of E_r and E_t to E_i yield the normal Fresnel coefficients, but in all other cases, the coefficients will clearly be a more complicated function of $A(\beta)$ and the angles θ_i and θ_t.

Secondly, it should be observed that the boundary conditions require that the propagation constants β_i, β_r, and β_t be equal, which can only occur if the normal laws of reflection and refraction are obeyed by the beams [3.8].

Thirdly, although not explicitly considered here, if the energy flow (Poynting vector) is calculated for the region just beyond the interface ($z \lesssim \lambda$), it is found that the flow is non-zero, and that its magnitude and direction change across the region of incidence [3.8]. This is illustrated in Fig.3.6, from which it can be seen that the left side of the incident beam leaks energy into the *less dense* medium, the right side returns energy into the *more dense* medium, while the central part of the beam energy flows in parallel to the boundary. In other words, the reflection appears to be slightly less than total for the left side of the beam, and slightly more than total for the right side. As a result, the weight center of the reflected beam is again shifted, with a corresponding shift in the apparent point of reflection on the interface.

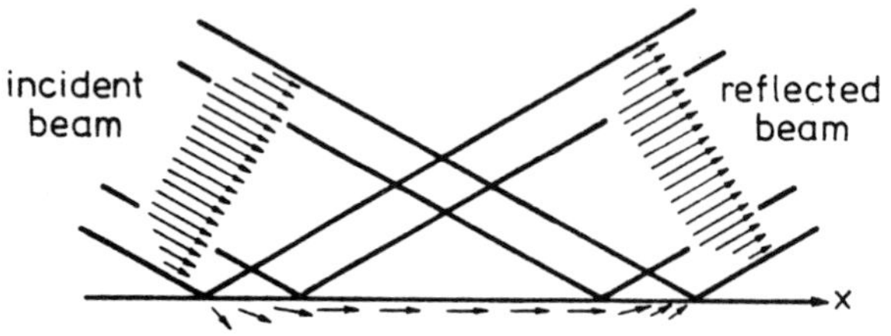

Fig. 3.6. Pictorial interpretation of local energy flows in the bounded beam and along the interface [3.8]

For an E-polarized beam incident at $\theta_i > \theta_{ic}$ (a singularity exists at $\theta_i = \theta_{ic}$), the lateral shift is approximately given by [3.8]:

$$X_\perp = \frac{\lambda}{\pi} \frac{\tan\theta_i(1-\sin^2\theta_i)}{(1-\sin^2\theta_{ic})\sqrt{\sin^2\theta_i-\sin^2\theta_{ic}}} , \qquad (3.16)$$

while the beam axis is displaced, as shown in Fig.3.7, by an amount

$$D_\perp = \frac{\lambda}{\pi} \frac{\sin\theta_i(1-\sin^2\theta_i)}{(1-\sin^2\theta_{ic})\sqrt{\sin^2\theta_i-\sin^2\theta_{ic}}} . \qquad (3.17)$$

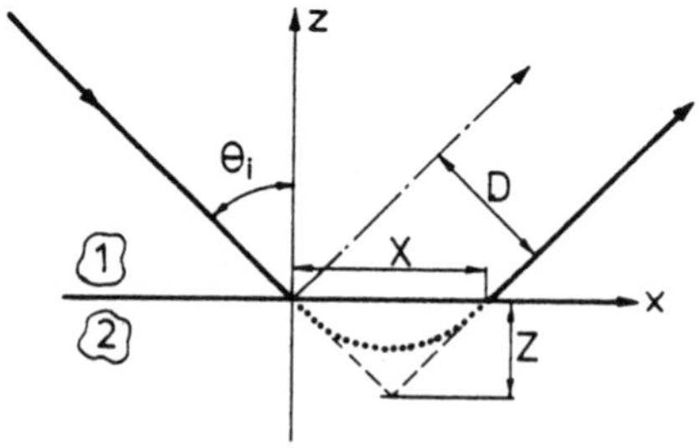

Fig. 3.7. Schematic representation of beam penetration into the less dense medium

The implication of (3.16) and (3.17) is that the reflection appears to originate from a *virtual* interface (Fig.3.7) that is located some distance from the interface.[1] The behaviour of the beam displacement is shown in Fig.3.8, while that of the penetration depth is shown in Fig.3.9. For the sake of completeness, curves for H polarized beams have also been included, and it can be seen that E and H polarized beams behave rather similarly. In fact, for large critical angles (as in optical fibers) beam displacements and penetrations become independent of polarization.

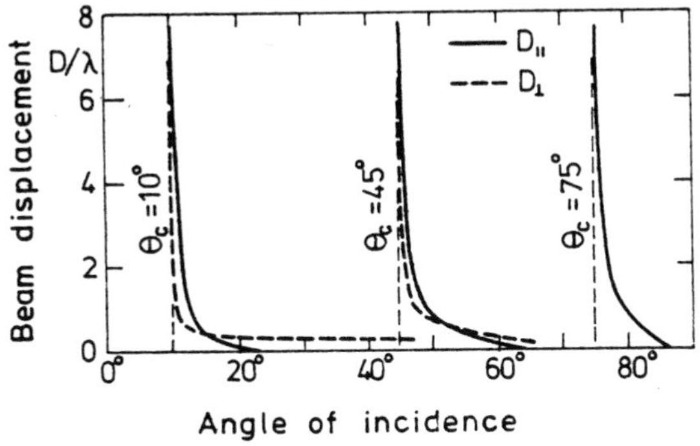

Fig. 3.8. Beam displacement D (normalized to λ) versus angle of incidence for E and H polarizations ($D_{\parallel}$ and $D_{\perp}$, respectively) [3.6]

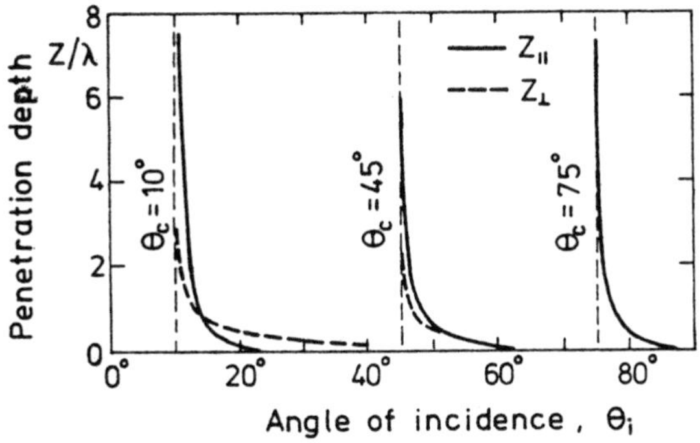

Fig. 3.9. Penetration depth Z (normalized to λ) versus angle of incidence for E and H polarizations ($Z_{\parallel}$ and $Z_{\perp}$, respectively) [3.6]

An evident feature of equations (3.16) and (3.17), as well as the corresponding H polarization formulae in [3.8], is that the beam shift appears to be independent of the beam width. At first sight, this seems strange, because

1 As a matter of historical interest, it may also be worth noting that the parabolic curve in Fig.3.7 shows the form of beam penetration that was predicted by NEWTON [3.9], based on his erroneous particulate theory of light.

on the basis of our previous plane-wave representation of the beam, we would expect that the narrower the beam, the wider would be its angular spectrum. As such, we would expect the beam width to influence all characteristics of reflection, including the beam shift. However, we must bear in mind that (3.16) and (3.17) are based on the assumption that the transverse beam amplitude changes slowly, so that we have inevitably been considering very wide beams (thousands of wavelengths).

The above wide beam restrsiction was removed by HOROWITZ and TAMIR [3.10], who showed that, for a Gaussian beam of the type

$$E_w = \frac{e^{-(\zeta/w)^2}}{w\sqrt{\pi}} , \tag{3.18}$$

the beam shift in the plane of the interface is approximately given by [3.2]

$$X_{\perp,\parallel} \simeq \frac{0.408\, mn^{\frac{1}{2}}}{(n^2-1)^{3/4}} \sqrt{w\lambda} \qquad \text{for } \theta_i \approx \theta_{ic} , \tag{3.19}$$

where w is half the beam waist, as defined by (3.18), and

$$m = \begin{cases} n^2 & \text{for H polarization} \\ 1 & \text{for E polarization.} \end{cases}$$

Equation (3.19) represents the limiting case ($\theta_i = \theta_{ic}$) of a general expression in [3.10] that reduces to (3.16) both for very wide beams and for θ_i sufficiently larger than θ_{ic}. For narrow beams ($w \sim 1000\ \lambda$), (3.19) states that the beam shift is proportional to $w^{\frac{1}{2}}$. This result is consistent with our earlier qualitative Fourier-optical approach, because when the central component of the beam is incident critically, about half of the beam power will be concentrated within the region defined by $\theta_i < \theta_{ic}$ and will be lost. As a result, the wider is the beam, the further away will be its weight center from the axis of geometrical reflection.

So far, we have been considering the time-averaged situation at the optical interface, and we should now examine the energy flow during one period of oscillation. To simplify the discussion, we will restrict ourselves to plane waves and assume that the following well-known phase-shift relations apply [3.1]

$$\tan\frac{\delta_\perp^r}{2} = \tan\delta_\perp^t = \frac{(n_1^2 \sin^2\theta_i - n_2^2)^{\frac{1}{2}}}{n_1 \cos\theta_i} , \tag{3.20}$$

$$\tan \frac{\delta_{\parallel}^r}{2} = \tan\delta_{\parallel}^t = \frac{n_1(n_1^2 \sin^2\theta_i - n_2^2)^{\frac{1}{2}}}{n_2^2 \cos\theta_i} \quad , \tag{3.21}$$

where δ^r and δ^t denote the phase changes upon reflection and transmission, respectively.

As shown by MAHAN and BITTERLI [3.11], these formulae have a significant influence on the instantaneous field values of all waves at the interface, because all boundary conditions for the fields, as well as the energy flow continuity condition, must be simultaneously satisfied at any given time. MAHAN and BITTERLI have calculated the incident, reflected, and refracted energy flows at the interface (for E polarization) with the results shown in Fig.3.10. The figure shows the instantaneous x and z components of the Poynting vectors of all three waves, for $n_1 = 1.5$, $n_2 = 1.0$, and $\theta_i = 45°$. These values of n_1 and n_2 yield a critical angle of 41.8°, so that incidence is near critical but very definitely in the total internal reflection regime.

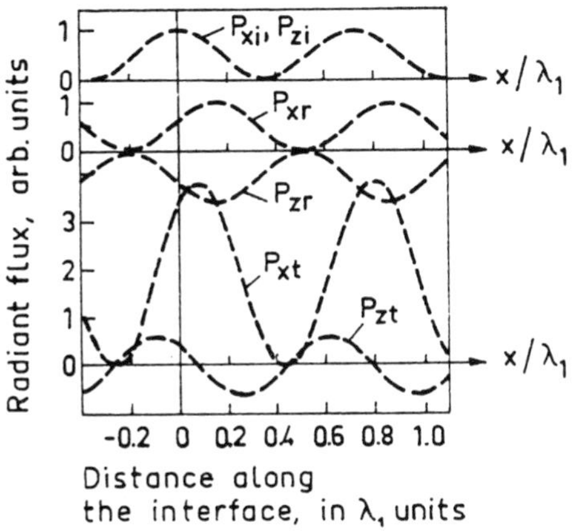

Fig. 3.10. Incident, reflected, and transmitted components of E polarized radiation [3.11]

First, note that the x and z components (P_{xi} and P_{zi}) of the incident wave are equal, and always positive for $\theta_i = 45°$. Their period about the x axis is $\lambda_1/(2 \sin\theta_i)$, implying that the distribution moves along the x axis at a velocity $v = c/(n_1 \sin\theta_i)$. The reflected components are similar to each other, but of opposite sign. They are also shifted to the right, by an amount $\lambda\delta_{\perp}^r/(2\pi \sin\theta_i)$, due to phase shift upon reflection.

Secondly, observe that there exist some points along the x axis (or some moments in time) at which $P_{zr} = P_{xr} = 0$, but $P_{zi} = P_{xi} \neq 0$. This surprising result states that only refraction occurs at such moments but no reflection. Similarly, there exist other times when we have no incident power, and still others when we have no refracted power. We are, of course, not very surprised that the incident power goes through zeroes since the incident wave is sinusoidal. However, it is perhaps harder to realize that the reflected and refracted waves do not synchronously follow the incident one. Note also that the x and z components of the refracted wave are different in amplitude and phase, so that the angle of energy penetration into the less dense medium varies with time. This effect is shown by Fig.3.11, which further shows that all "refracted" energy is returned to the denser medium. In other words, we have total internal reflection as we would expect, but the energy also dives in and out of the less dense medium. We now see that the depth of penetration is not "virtual" as would appear from the qualitative arguments used to explain the Goos-Hänchen shift, but is very real and physical. MAHAN and BITTERLI also gave the time averaged power in the less dense medium (as a function of distance from the interface) to be

$$\langle P_t \rangle = \langle P_{xt} \rangle \sim \exp\left[-\frac{4\pi n_1}{\lambda} z\left(\sin^2\theta_i - \frac{n_2^2}{n_1^2}\right)^{\frac{1}{2}}\right] \quad . \tag{3.22}$$

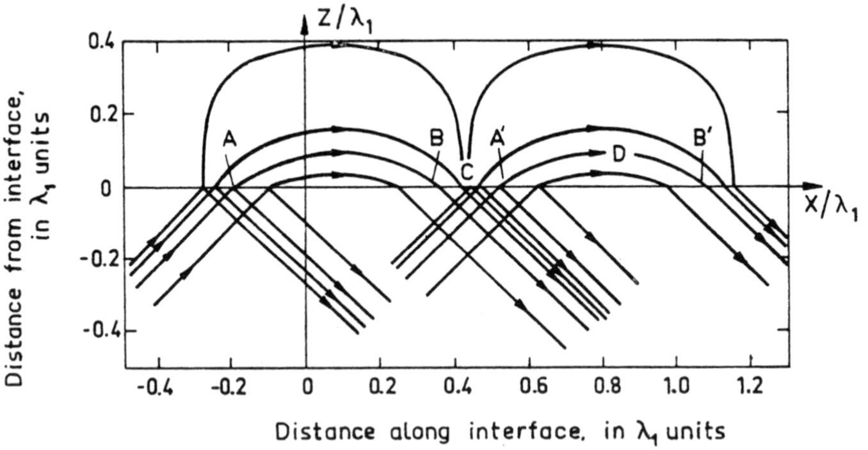

Fig. 3.11. Instantaneous energy flow lines at total internal reflection [3.11]

We see that the energy penetrates exponentially into the less dense medium, making it possible for a third medium to extract power: the nearer to the interface, the more the extracted power. This is the origin of the phenomenon known as frustrated total internal reflection.

In the foregoing, we have basically discussed power flow along the interface at a given instant in time. However, we should also look at what happens at a given point near the interface as a function of time. This is equivalent to moving the pictures in Figs.3.10 and 3.11 past the observation point, at the previously mentioned velocity of $v = c/(n_1 \sin\theta_i)$. Thus, at point D, and at the time frozen in Fig.3.11, the energy flow is parallel to the interface. As the picture moves from left to right, the direction of energy flow first changes from parallel to perpendicular, and then back again to parallel. This is accompanied by a change in the mangitude of the energy, from maximum when flowing parallel to the interface, to zero when perpendicular to the interface.

Finally, we should return to the Fourier-optical description, and consider the behavior of a beam that is incident at a substantially greater angle than critical, and which contains a negligible number of plane-wave components with $\theta_i < \theta_{ic}$. According to our earlier arguments, it might seem that, in this case, no Goos-Hänchen shift is possible. However, in view of the phase shifts discussed above, we can say that the phase angle of each reflected plane-wave component will be slightly modified. As a result, beam shift due to phase distortion becomes possible, but from (3.16) and (3.17) we know that any such shift must be small in comparison to the one obtained for critical or near-critical incidence.

3.2 Reflection Phenomena at a Curved, Blurred, or Periodically Modulated Interface Between Two Media

In the previous section, we discussed total internal reflection at a plane interface. We would now like to consider the influence of the interface shape and sharpness upon the reflection characteristics. Of course, the shape of the interface may be non-planar either deliberately, as in fibers and in surface-corrugated thin film couplers, or it may be affected by technological and environmental limitations (diffusion, bends, etc.). However, we shall consider the latter limitations to be of second order and confine ourselves to well behaved deviations from the plane interface.

Let us start with the situation in which a plane wave is incident at an interface that is curved into the plane of incidence, with a radius of curvature ρ (Fig.3.12). In this case, the local angle of incidence is different

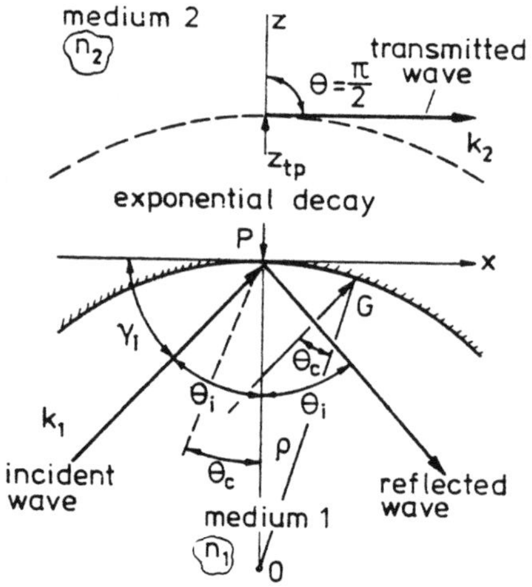

Fig. 3.12. Pictorial representation of tunnelling due to reflection at a curved interface between two non-absorbing media [3.12]

at different points along the interface. If we define the angle at the point P (x = z = 0) to be θ_i, such that θ_i is greater than the critical angle θ_{ic}, then it can be seen that the angle of incidence decreases to the right of P until, at some point G, it becomes less than critical. Thus, according to the results of Sect.3.1, it would seem that *all* points to the right of G should give rise to a refracted wave. However, it can be shown [3.12] that, close to the curved interface (in the less dense medium), the z component of the propagation vector becomes imaginary. In other words, for a plane wave given by:

$$E_2 \sim \exp\left[j(k_{2x}x + k_{2z}z)\right] \quad , \tag{3.23}$$

where

$$k_{2x}^2 + k_{2z}^2 = k_2^2 \quad , \tag{3.24}$$

we find that

$$k_{2x} > k_2 \quad , \tag{3.25}$$

or that

$$k_{2z}^2 = k_2^2 - k_{2x}^2 < 0 \quad . \tag{3.26}$$

As a result, the field in (3.23) decays exponentially along the normal into the less dense medium. However, as shown in [3.12], beyond a certain distance

z_{tp}, k_{2z} becomes real and a radiating wave is obtained for

$$z > z_{tp} = \rho[(\sin\theta_i/\sin\theta_{ic}) - 1] \quad . \tag{3.27}$$

This effect of plane-wave solutions on both sides of a region of exponential decay is known as "tunnelling", by analogy to its namesake in quantum mechanics.

Let us now specify our two media to be slightly absorbent, so that their indices of refraction assume the form

$$\tilde{n}_1 = n_{1r} + jn_{1i} \quad , \tag{3.28}$$

$$\tilde{n}_2 = n_{2r} + jn_{2i} \quad . \tag{3.29}$$

In this case, the question of the relative influence of absorption and tunnelling becomes interesting. These two effects can be compared by using a two-part power transmission coefficient [3.13]

$$T \approx T_F + T_C \quad , \tag{3.30}$$

where T_F is the coefficient due to absorption (calculated for a plane interface), T_C is due to tunnelling, and T is the generalized Fresnel coefficient. It is found that (3.30) holds only when

$$\theta_i \geqq \theta_C \quad , \tag{3.31}$$

and

$$|(k_1\rho \sin\theta_C)^{2/3} (n_{1i}/n_{1r} - n_{2i}/n_{2r})| \ll 1 \quad . \tag{3.32}$$

The latter condition ensures that

$$T_C \gg T_F \quad , \qquad \text{when } \theta_i = \theta_C \quad . \tag{3.33}$$

The angle θ_C differs from the normal critical angle θ_{ic}, and is defined by

$$\sin\theta_C = n_{2r}/n_{1r} \quad , \tag{3.34}$$

such that for non-absorbing media ($n_{1i} = n_{2i} = 0$) we obtain $\theta_C = \theta_{ic}$.

Under these assumptions, the behaviour of the transmission coefficient can be analyzed, with the results summarized in Fig.3.13. The figure shows how the product $T\gamma_c$ [where $\gamma_c = (\pi/2) - \theta_c$] varies with $(\gamma_c^2 - \gamma_i^2)$, for nearly equal indices of refraction. The solid curve is for a slightly absorbent $(n_{2i}/n_{2r} \approx 10^{-8})$ less dense medium, while the broken curves are for the lossless case $(n_{2i} = 0)$. All curves have been plotted for a large variety of radii of curvature, and we see that, for ρ less than a few centimeters, the broken and solid curves coincide. Thus, it is clear that in these cases curvature is predominantly responsible for power transfer through the interface and $T \approx T_C$. Conversely, for larger curvatures e.g. $\rho = 1$ m $(k_1\rho \sim 10^7)$, the power transfer is almost independent of the angle of incidence, implying that $T \approx T_F$ and that absorption is the dominant mechanism of power transfer. Note, however, that for non-absorbent media, even a small amount of curvature will give rise to a non-zero transmission coefficient, at least for a small range of angles near critical.

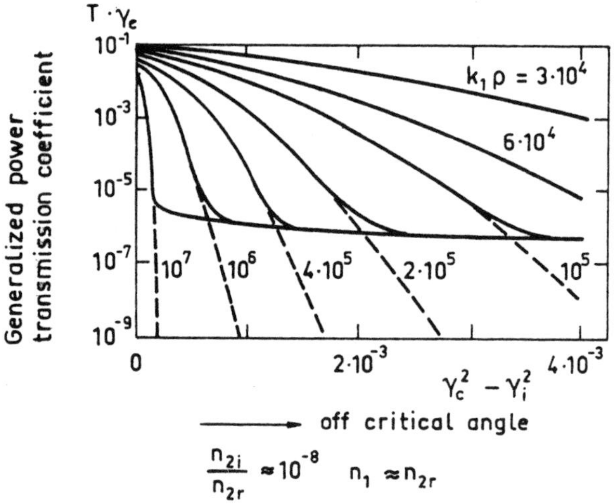

Fig. 3.13. Power transmission coefficients against off-critical angle for a slightly absorbent less dense medium (——), and a non-absorbent less-dense medium (----) [3.13]

Let us next consider the behaviour of a plane wave at a cylindrical interface of radius ρ, such that the core of the cylinder has a refractive index n_1, and the surrounding region has a lower index n_2. This is equivalent to extending the curved surface of Fig.3.12 into a cylinder. We can now divide the continuum of incident waves into three groups, each with its own distinct behaviour [3.14]. The first group is formed by waves that are incident at $\theta_x \leqq \pi/2 - \theta_c$, and are totally reflected and "trapped" by the cylinder, while the second group contains waves that are incident at $\theta_x > \pi/2 - \theta_c$ and are classically refracted. The propagation vectors of these waves define two half cones, one with a semi-vertical angle $\pi/2 - \theta_c$ about the x-axis, and the other with a semi-vertical angle of θ_c about the normal (Fig.3.14). The first cone defines the angular space within which all waves are reflected, while the second defines the space within which all waves are refracted. The remaining radiation falls outside these two angular spaces and forms the third group of partly reflected and partly tunnelled waves.

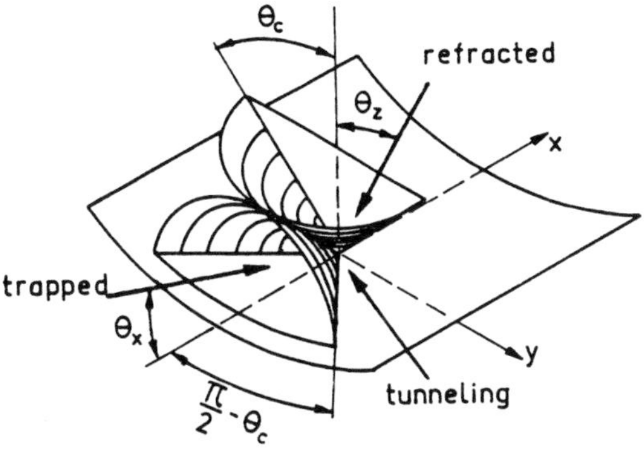

Fig. 3.14. Different angular zones in the vicinity of the point of incidence defining the behaviour of a plane wave with a given propagation vector [3.14]

As we have already seen, the width z_{tp} of the region of exponential decay, will depend on the effective radius of curvature "seen" by the incident wave. Due to the inclination of the plane of incidence with respect to the axis of the cylinder, the interface seen by the wave is the section of an ellipse, with a local curvature of $\rho_i = \rho \sin^2\theta_z/\cos^2\theta_y$. Thus, for meridional waves (for which $\rho_i = \infty$), we can only have reflection or refraction, but not tunnelling. The line along which the two semi-cones meet in Fig.3.14 depicts this situation. Conversely, in the cross-sectional plane of the cylinder in which $\rho_i = \rho$, we obtain refraction for $\theta_z < \theta_c$ and tunnelling for $\theta_z > \theta_c$, but no total reflection is possible. All other cases represent the intermediate situation, and each of the three possibilities (reflection, refraction, and tunnelling) must be considered.

Up to this point, we have assumed that the interface is ideally sharp, i.e. the change in the index is a step function. However, the transition from one medium to another is not usually of this type but is blurred in some way. We will once again restrict ourselves to a plane interface, and consider the index change to occur either in a series of small steps, or monotonically, as

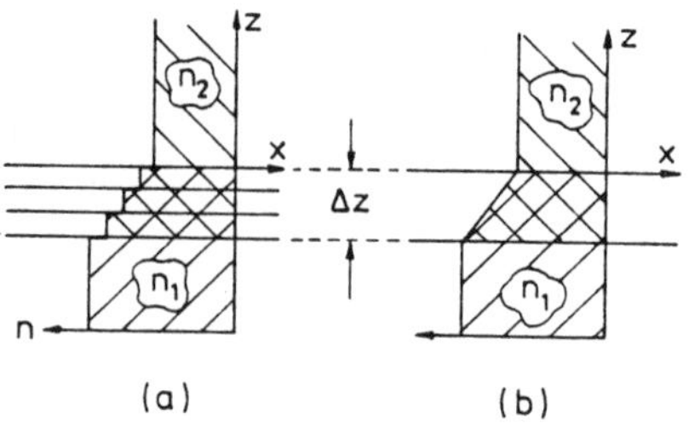

Fig. 3.15. Representation of the refractive index profile along the normal to the interface: (a) stepwise graded; (b) monotonically graded

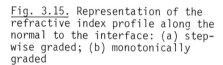

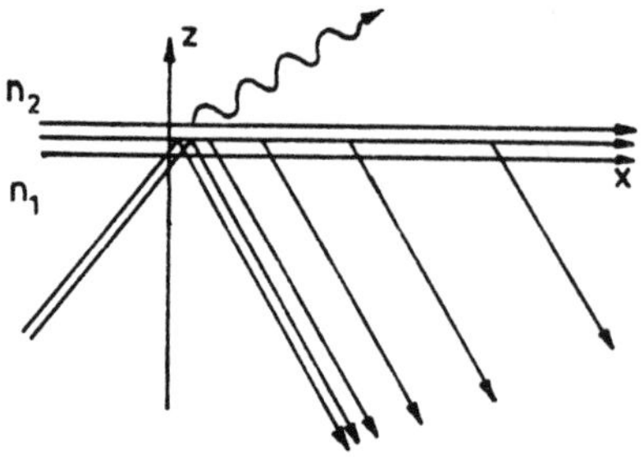

Fig. 3.16. Pictorial representation of the behaviour of a bounded beam at a multi-layered interface

shown in Figs.3.15a and b, respectively. Let us first consider the former case, which in fact corresponds to having a multi-layered or stratified boundary region [3.15]. Then, referring to Fig.3.16, we can make the following observations:

1) If a bounded beam is incident upon the multilayered boundary, the first interface will reflect those Fourier components that have their propagation vectors at angles greater than the critical. The rest of the components will be refracted.

2) The refracted components will now be incident on the second interface. This interface will have a slightly smaller critical angle, and will be able to reflect some of the incident components. The reflected components will be once again incident on the first interface, but will now undergo *partial* reflection and refraction, since the lower medium is more dense. The partially reflected components will in turn be totally reflected by the second interface, and will again be incident on the first interface. The result is that, although a part of the originally incident energy propagates in the layer between the first and second interfaces, the energy is gradually returned to the lower medium and contributes to the reflected beam.

3) Similarly, every layer within the stratified region propagates some energy in the x direction but gradually returns it to the lower medium. Consequently, the profile of the reflected beam will be different from that of the incident beam. In other words, the weight center of the reflected beam

will be shifted. Note, however, that the mechanism is somewhat different from the one associated with the Goos-Hänchen shift (Sect.3.1).

4) The thinner is each layer, the more will be the number of partial reflections per unit distance, and the more rapidly will the energy be returned to the main reflected beam.

5) For each layer, the medium immediately below its lower interface is always more dense. Thus, the layer can never quite form a true waveguide.

The case of diffuse boundaries of the type depicted in Fig.3.15b have been studied by KOZAKI and co-workers [3.16-18], who have used an incident beam of Gaussian profile, and have presented numerical results for the case of a linearly changing refractive index (or permittivity). As in the stratified case, they show that the reflected beam undergoes distortion. However, unlike the stratified case, the Gaussian beam suffers a significant compression as it moves through the transition layer (Fig.3.17), the compression being strongest at the turning point (Fig.3.18).

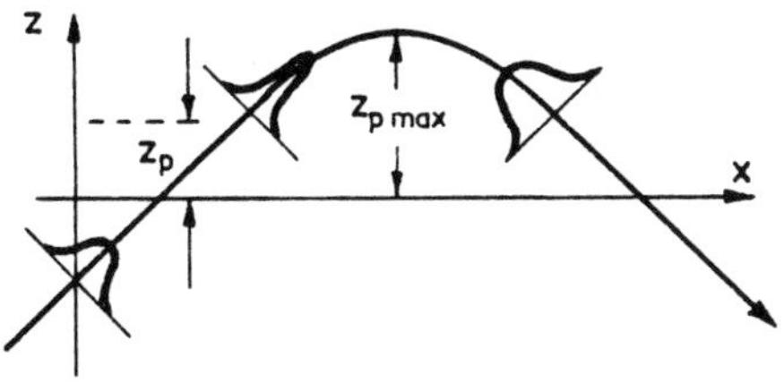

▲ Fig. 3.17. Behaviour of a bounded beam at a diffuse boundary

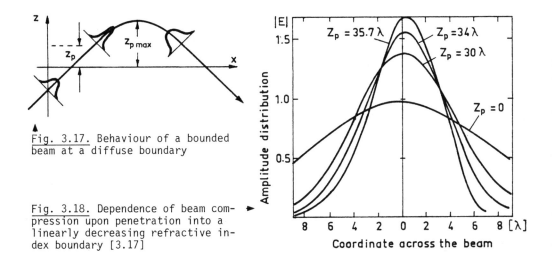

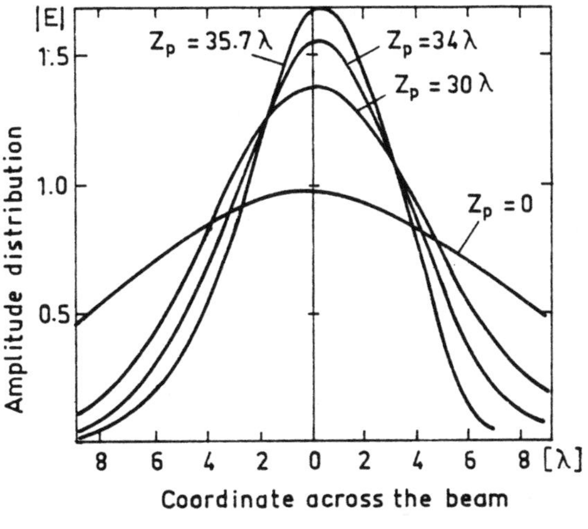

Fig. 3.18. Dependence of beam compression upon penetration into a linearly decreasing refractive index boundary [3.17] ▶

The last situation to be considered in this section is wave behavior at a periodically modulated interface. The modulation may be either deliberate (as in couplers) or it may be caused by inevitable interface imperfection. When deliberate, it will be due to a periodic change in some parameter of the interface. Thus, we may have surface corrugations, refractive index modulation, periodic metallization, etc. Whatever the form of modulation, a complete analysis of modulated interfaces is either hard or time-consuming. For example, in the case of surface corrugations, the local angle of incidence is periodically modulated, and the analytical satisfaction of boundary conditions becomes quite a problem. Another example is the refractive-index modulation of a thick intermediate layer. This case has been studied by KOGELNIK [3.19,20] via a wave-coupling analysis that is limited to orders obeying the Bragg condition. Perhaps the simplest case is that of thin metal stripes deposited on the interface, but this is also rather difficult to analyze because of strong absorption effects [3.21]. However, if the spatial period of this metallic grating is much greater than the wavelength, the reflection phenomenon can be treated in terms of diffraction at a phase grating [3.22,23].

We will restrict ourselves to this simplest of all cases, in the hope that it will provide some qualitative insight into the behaviour of all types of modulated interfaces. Let us first define the complex reflection coefficients $\tilde{R}_1$ and $\tilde{R}_2$ for the metallized and non-metallized parts, respectively;

$$\left.\begin{aligned} \tilde{R}_1 &= R_1 \exp(j\delta_1) \quad , \\ \tilde{R}_2 &= R_2 \exp(j\delta_2) \quad , \end{aligned}\right\} \tag{3.35}$$

where R_1 and R_2 are the amplitudes of the reflection coefficients, and δ_1 and δ_2 are the phases. (All these coefficients are rapidly changing functions of the angle of incidence). Then, based on Fourier optics [3.24], the diffraction efficiencies of our grating can be shown to be [3.23]:

$$\begin{aligned} \eta_0 &= I_r^0/I_i = [R_1^2 + R_2^2 + 2R_1R_2\cos(\delta_2 - \delta_1)] \quad , \\ \eta_{\pm 1} &= I_r^{\pm 1}/I_i = 4\eta_0/\pi^2 \quad , \end{aligned} \tag{3.36}$$

where I_i, I_r^0, and $I_r^{\pm 1}$ are the intensities of the incident wave, and zero and first order diffracted waves, respectively.

Figure 3.19 shows the variation of diffraction efficiency as a function of the incidence angle θ_i. Chromium metal stripes and an air-glass interface have been assumed. For the former, tabulated values of R_1 and δ_1 have been used, while for the latter R_2 and δ_2 have been obtained using Fresnel's formulae. The main feature of these curves is a strong anomaly in diffraction efficiency near the critical angle, and a comparatively high efficiency for E-polarized light ($\eta_{+1} \sim 30\%$). The experiments carried out by UDOJEV and OVCHINNIKOV [3.23] show close agreement with the theoretical predictions (Fig.3.20).

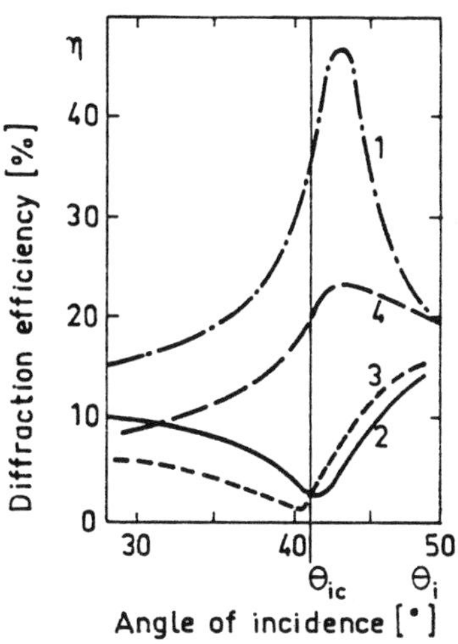

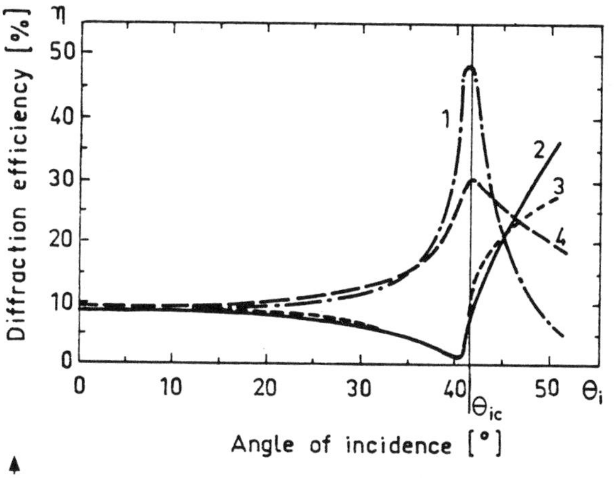

▲ Fig. 3.20. Experimental results for the cases in Fig.3.19 [3.23]

▲ Fig. 3.19. Theoretical diffraction efficiencies of a metal grating on a quartz substrate [3.23]. (1) η_0 for H polarized wave; (2) η_0 for E polarized wave; (3) η_{-1} for H polarized wave; (4) η_{-1} for E polarized wave

In summary, we can say that, whenever a planar step-index interface between two media is disturbed (curved, blurred, or periodically modulated), strong changes occur in the laws normally associated with total internal reflection. Not only do we have energy leakage into the less dense medium, but for bounded beams the profile of the reflected beam may be strongly deformed, or even split up into diffraction orders. Such changes in the propagation direction of the radiation produce the phenomenon of "mode coupling" whereby the energy in a given waveguide mode can be transferred to other modes, often with detrimental consequences.

3.3 Planar Step-Index Waveguides

The basic structure of a planar step-index waveguide is shown in Fig.3.21a. The waveguide consists of a semi-infinite substrate of refractive index n_s, a film of refractive index n_f deposited on top of it, and a second semi-infinite medium of refractive index n_c covering the film. The film is thus bounded by two interfaces, between which a suitably incident wave can propagate by means of total internal reflection. We will assume that the film is denser than both the substrate and the cover. Such a waveguide is termed symmetric if the refractive indices of the substrate and cover are equal, and asymmetric if the index of the substrate is higher than that of the cover.

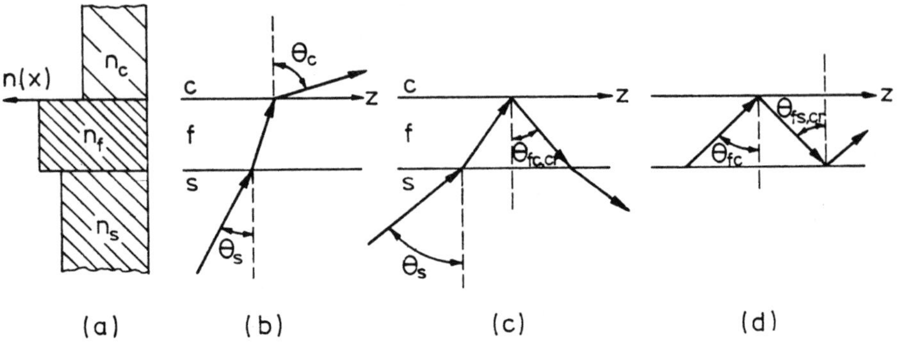

Fig. 3.21. (a) Refractive index profile of step-index waveguide; (b) Radiation mode escaping via the cover; (c) Radiation mode escaping via the substrate; (d) A propagating mode

Step-index waveguides (we shall omit the term planar when there is no ambiguity) have been rather thoroughly studied, and rigorous mathematical treatments have been given, amongst others, by LOTSCH [3.25], KOGELNIK [3.26], BARNOSKI [3.27], and MARCUSE [3.28]. In line with some of these authors, we shall change the previously used orientation of the x and z axes, in order to follow the tradition of using the z axis as the direction of energy propagation. As before, the y axis will be normal to the plane of all drawings, and all fields will be independent of this coordinate.

Let us then examine the behaviour of a plane wave, which strikes the lower interface of the film. If the angle of incidence at the upper interface remains less than critical (Fig.3.21b), then a small part of the wave's energy will be reflected, but most of it will be refracted. The reflected portion will rapidly disappear from the film after multiple partial reflection. If the angle of incidence at the upper interface is made slightly more than critical (Fig.3.21c), then total internal reflection occurs. But if the guide is asymmetric, then the critical angle at the lower interface will be larger than that at the upper, and the wave will not be confined to the film. However, if the angle of incidence is increased still further (Fig.3.21d), such that it becomes greater than the critical angle of the lower interface, then the wave will be confined and will propagate in the z direction. We see that although total internal reflection, with all its attendant features, is the main effect responsible for propagation, the existence of two interfaces results in a form of resonator, and we would expect the wave behaviour to be modified in some way.

In order to understand this behaviour, we propose to use a simple graphical technique which illustrates the underlying physics. Thus, with reference to Fig.3.22, suppose that a plane wave is incident at the lower interface of the guiding film such that total internal reflection occurs at the upper interface, but not at the lower. Let the solid lines in the figure represent instantaneous positions of the maxima of the wave, and the dashed lines the instantaneous positions of anti-maxima (i.e. negative maxima). Then, the separation between maxima (or anti-maxima) is λ/n_s in the substrate, and λ/n_f in the film, with the latter being less than the former. (N.B. λ is the free space wavelength). We note that, in order to achieve amplitude continuity across the lower boundary, the wave must change direction in the film (Snell's law).

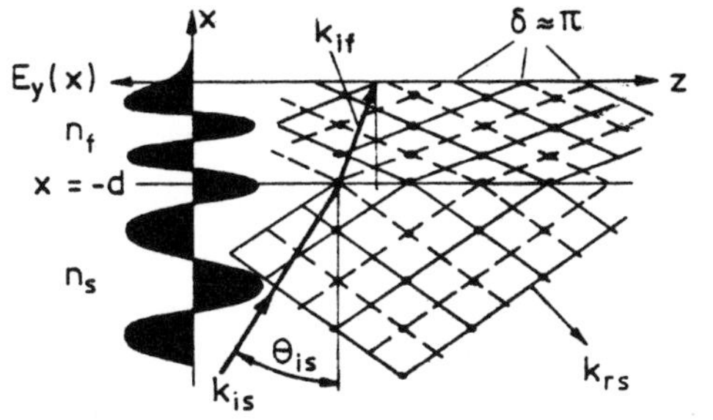

Fig. 3.22. Graphical illustration of the standing-wave pattern for total internal reflection at the upper interface only

The next step is to refer to Fig.3.23, which is essentially a plot of (3.20), and shows half the phase shift between incident and reflected waves. For the sake of simplicity, we will assume that the half angle in our case is 90° or that the phase shift is 180°. (This assumption is not realistic but will serve to illustrate the technique). To satisfy this shift requirement the intersections of solid and dashed lines must coincide with the interface, since solid lines (maxima) and dashed lines (anti-maxima) are 180° out of phase. This fixes the relative positions of all reflected wavefronts in the film and substrate. (Upon crossing the lower interface, the wave will, of course, once more have to satisfy Snell's law).

We can see in Fig. 3.22 that certain planes parallel to the interfaces contain intersections of solid lines at some points, and dashed lines at others. At these points, we expect constructive interference to yield maxi-

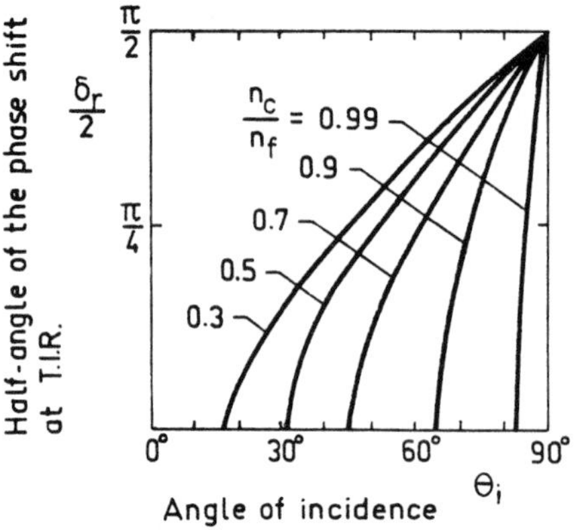

Fig. 3.23. Dependence of phase shift upon the angle of incidence

ma or anti-maxima of the electric field. Similarly, we can find planes which contain intersections of solid and dashed lines, and in which we have destructive interference (zero fields). All other planes represent some intermediate situation. In the time domain, this instantaneous distribution moves in the z direction (cf. Sect.3.1) with a velocity given by

$$v = c/n_s \sin\theta_{is} = c/n_f \sin\theta_{if} \quad .$$

However, observe that the aforementioned intersection points remain in their own planes and preserve their separations. Thus, the field distribution in the x direction must remain stationary. In other words, we will have a standing wave pattern in the x direction as shown on the left side of Fig.3.22. We reach the conclusion that, by using this graphical method, we can estimate the time-averaged energy distribution along the x axis by simply determining the number and position of planes containing field maxima, anti-maxima, and zeroes. Note that the exponential decay of the field beyond $x > 0$ cannot be predicted using this method, but is something which we expect from the results of previous sections [see (3.22)].

The graphical technique can, of course, also be applied to the situation in which the film confines the wave, and total internal reflection occurs at both interfaces. As before, all possible field distributions (across the film) can be found by drawing the instantaneous positions of incident and reflected wavefronts. However, in this situation we must allow for phase shifts at both interfaces, as well as restrict the angle of incidence to values which are greater than or equal to the larger critical angle (lower interface). In our simplified treatment we again assume a phase-shift of π at each interface, but whatever the phase shift requirement, it can only be fulfilled for discrete angles of incidence[2]. After establishing the "allowed" angles, the field distribution across the film can, as before, be estimated by determining the positions of planes which contain maxima, anti-maxima, and zeroes. We have drawn two examples in Fig.3.24: the upper shows the lowest order mode (largest allowed angle of incidence) and the lower a higher order mode. By using this technique, we therefore can roughly determine allowed modes and their angles. The rigorous approach would, of course, be to solve Maxwell's equations for the fields in the film. Then, matching solutions across the interfaces would yield the so-called eigenvalue equation, which would then define the allowed modes.

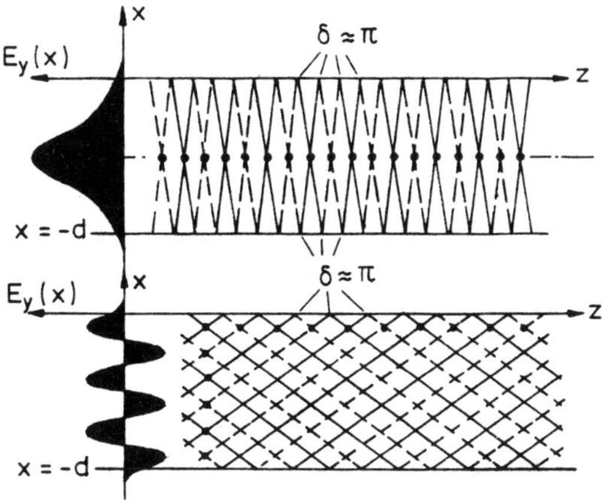

Fig. 3.24. Illustration of standing wave patterns of two propagating modes

Fig. 3.25. Ray trace for determining the phase difference between two rays that are required to interfere constructively

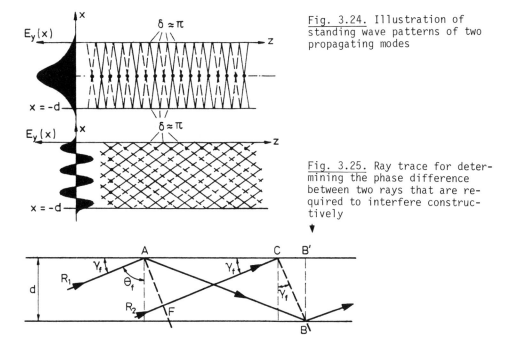

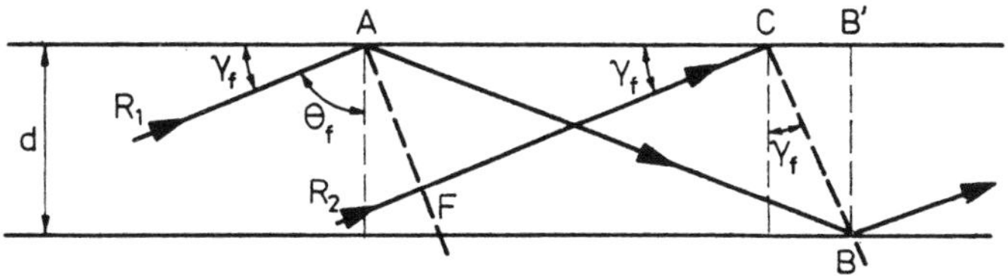

However, for our purposes, it is sufficient to adopt an approach used e.g. by UNGER [3.29]. His treatment gives the same transverse resonance condition as the more rigorous analysis, but is much simpler. Consider therefore the situation shown in Fig.3.25, in which two "rays" R_1 and R_2 are incident at points A and C. The point C has been chosen such that the plane

2 We suggest that the reader perform a convincing experiment using three pieces of transparent plastic, on two of which are drawn "solid" and "dashed" wavefront positions, and on the third the position of the interfaces. Mutually positioning the three pieces to obtain the required phase shift, it will immediately be noticed that only discrete angles are allowed.

wavefront just after reflection at B goes through it. Now, R_1 will suffer from total internal reflection phase shifts, say δ_{fc} and δ_{fs}, at A and B, as well as travel a distance $\overline{AB}$ which is longer than the distance $\overline{FC}$ travelled by R_2. This extra distance causes the temporal fields associated with R_1 to be delayed, say by τ, so that in the frequency domain (Fourier transform of time) the fields must be multiplied by a factor $\exp(-j\omega\tau)$. Thus, the additional distance travelled is equivalent to a phase shift $(-\omega\tau)$. The algebraic sum of these various phase shifts must be an integer multiple of 2π in order to have a self-consistent solution. Violation of this condition implies that the wavefront arriving at C is not in phase with the wavefront just reflected from B, so that the wave would then interfere destructively with itself, and would ultimately die out!

Looking now at Fig.3.25, we note that the path difference of interest to us is $\overline{AB} - \overline{FC} = \ell$. Then, the phase shift, say Φ_ℓ, is given by

$$\Phi_\ell = -\omega\tau = \frac{-2\pi n}{\lambda} \cdot \ell = -k_f \cdot \ell \quad . \tag{3.37}$$

Furthermore, simple geometrical considerations yield

$$\overline{AB} = d/\sin\gamma_f , \tag{3.38}$$

$$\overline{FC} = \overline{AC}\cos\gamma_f \quad , \tag{3.39}$$

and $$\overline{AC} = d/\tan\gamma_f - d\tan\gamma_f \quad . \tag{3.40}$$

Use of these equations yields the phase shift due to path difference:

$$\phi_\ell = -2db_f \tag{3.41}$$

where

$$b_f = k_f \cos\theta_f \tag{3.42}$$

represents the x component of the wave vector k_f. Equating the sum of the phase shifts to $-2\pi m$, we obtain the so called eigenvalue equation for propagating modes

$$2b_f d = \delta_{fc} + \delta_{fs} + 2\pi m \quad , \quad \text{where} \quad m = 0, 1, 2, \ldots \quad . \tag{3.43}$$

Thus, we see once again that only discrete angles of incidence are allowed. The application range of this important equation can be determined by considering that, in general, the wave in each region is of the form

$$E \sim \exp[j(k_x x - \beta z)] \quad , \tag{3.44}$$

where β is the axial propagation constant. The value of β must be the same in all three regions since the axial velocity must be uniform. Furthermore, the above form is applicable for propagating solutions in the film, but for decaying behaviour in the substrate and cover k_x must become imaginary. Thus, using (3.6), we can say that β must exceed k_x but be less than k_f. In other words, for propagating modes

$$\frac{2\pi n_f}{\lambda} \geqq \beta > 2\pi n_s/\lambda \quad . \tag{3.45}$$

The resolved propagation vectors now become

$$\left.\begin{aligned} b_f^2 &= k_f^2 - \beta^2 \\ b_s^2 &= \beta^2 - k_s^2 \\ b_c^2 &= \beta^2 - k_c^2 \quad . \end{aligned}\right\} \tag{3.46}$$

Thus, we see that (3.45) gives us the range of β, and (3.46) provides us with the corresponding range of b_f. Within this range, we can then solve (3.43), for all allowed values of m. [Each value of m represents a specific mode with a specific standing wave pattern (Fig.3.24)].

Returning now to the eigenvalue equation, we note that, in practice, it is usually solved numerically. However, in order to obtain some further information about the waveguide, we will solve the equation graphically, but only for the lowest order mode (m = 0). In this case, (3.43) reduces to

$$2\pi n_f d\,(\cos\theta_f)/\lambda = (\delta_{fc} + \delta_{fs})/2 \quad . \tag{3.47}$$

The solution of this equation is shown in Fig.3.26. The left side of (3.47) has been drawn for two different thicknesses. The upper curve corresponds to the thicker waveguide ($d_2 > d_1$), and the lower curve to the thinner.

In the case of a symmetric waveguide ($n_s = n_c$, $\delta_{fs} = \delta_{fc}$), the right side of (3.47) becomes δ_{fs}, and we see that, for every value of waveguide thickness, the intersections of δ_{fs} with the left side of (3.47) correspond to angles of incidence which are greater than the critical value $\theta_{fs,cr}$. Thus, a symmetric waveguide of arbitrary thickness can always support the lowest-order mode.

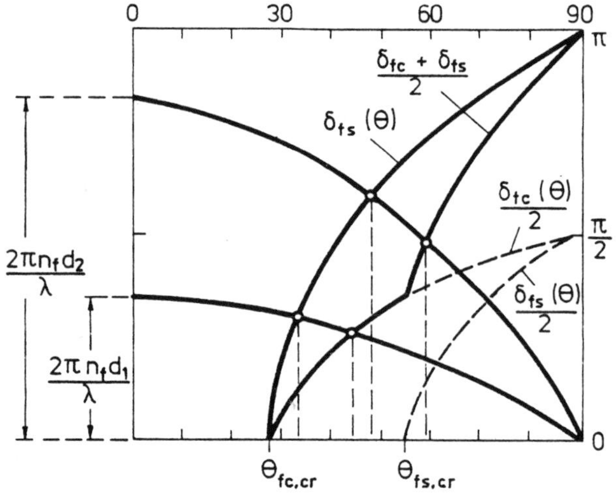

Fig. 3.26. Graphical solution of the eigenvalue equation for two different thicknesses (d_1 and d_2) and symmetrical (——) and asymmetrical (----) cases

The same is not true for asymmetric waveguides. In this case, the intersection of the left- and right-hand sides of (3.47), for $d = d_1$, corresponds to an angle of incidence which is less than the critical value, $\theta_{fs,cr}$. Hence, total internal reflection cannot take place at the lower interface, and there can be no guided mode. However, for $d = d_2$, the intersection corresponds to an angle of incidence which is greater than the critical angles of both interfaces. Thus, this guide will not be in a "cut-off" condition and will be able to support propagation.

To obtain some idea about the number of modes that a waveguide can support, we must solve (3.43) for the given waveguide. Figure 3.27, which is the result of such a computation, shows the dependence of the so called mode index, $N = \beta/k$, upon the thickness of the waveguide, for the specific case of $n_f = 2.29$, $n_s = 1.5$, $n_c = 1.0$, and $\lambda = 1060$ nm [3.30]. The following observations can be made:

1) "Cut-off" occurs somewhere at $d_0 \leqq 0.1$ μm, i.e. the film cannot support any mode under these circumstances.

2) With increasing values of d, the number of propagating modes also increases. Note also the close similarity between TE and TM modes of the same index. (For mode definitions see Sect.3.5).

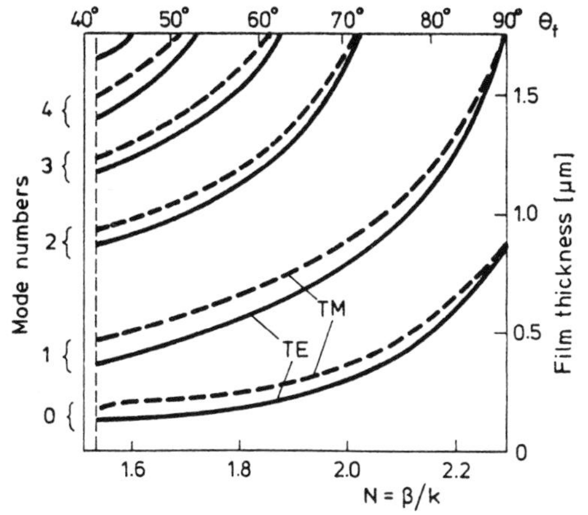

Fig. 3.27. Mode diagram for a step-index waveguide with the refractive indices given in the text [3.30]

3) For angles of incidence greater than the critical angle of the film-substrate interface ($\theta_{fs,cr} \approx 41°$), it is possible to excite all modes. This corresponds to a critical mode index of $N \approx 1.5$.

4) For angles of incidence less than this critical value, no waveguide modes can be excited. Note, however, that under these conditions substrate modes *can* exist.

5) Figure 3.27 can also be used for design purposes. For example, we can see that a waveguide with a film thickness of 1.5 μm will support 5 TE and 5 TM modes.

In concluding this section, we would like to stress that although we have demonstrated discrete modal behaviour for the simplest type of optical waveguide, it also arises in other more complicated structures. This will be further confirmed in Sections 3.4-6, but physically, modal behaviour is most easily demonstrated in the step-index-planar-waveguide case.

3.4 Planar Graded Index Waveguide

In the previous section, we discussed the case of a planar waveguide with sharp and well defined interfaces. In practice, such interfaces are frequently blurred due to diffusion, so that some of the effects discussed in Section 3.2

come into play. However, in contrast to Section 3.2, here we will assume that the whole film is blurred, and that its index is monotonically graded. Will such a film confine and guide light? The first part of this section is devoted to answering this question on the basis of a ray path analysis.

On the basis of the results of Section 3.3, we expect that our graded resonator should also force propagation in the form of discrete modes. In this case, however, it is difficult to predict simple phase shift relations for the general graded index profile, in order to obtain the eigenvalue equation. Thus, after the ray analysis we resort to Maxwell's equations, which for the graded guide can be solved by using the Wenzel-Kramers-Brillouin (WKB) approximation. Another reason for presenting this relatively complex analysis is that it will give us further insight into the behaviour of graded index fibers, and will provide the basis for an even more complex analysis.

We first derive the ray path by studying the differential geometry of a ray in a stratified film. Fig.3.28 shows two such stratifications of refractive index n_1 and n_2, with their interface located at x_2. A ray incident at an angle ψ_1 with respect to that interface undergoes refraction according to Snell's law, and emerges at an inclination ψ_2. The path lengths of the ray are s_1 and s_2 in n_1 and n_2, respectively. Thus, we can write

$$n_1 \cos\psi_1 = n_2 \cos\psi_2 = (n_1 + \Delta n) \cos(\psi_1 + \Delta\psi) \quad . \tag{3.48}$$

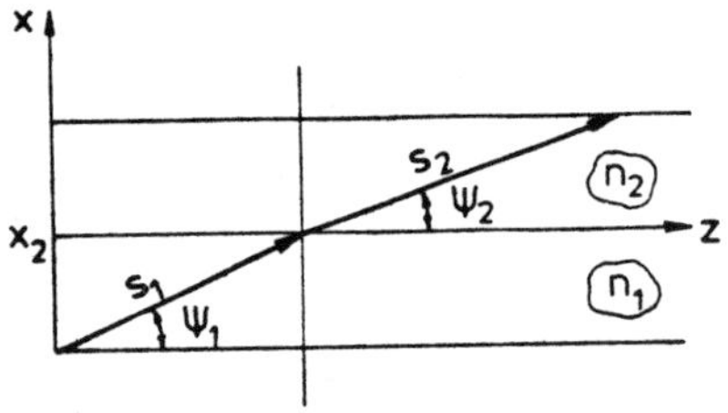

Fig. 3.28. Representation of the ray path in a thinly layered structure

Discarding terms of second or higher order in $\Delta\psi$, and taking the difference of $n_1 \cos\psi_1$ and $n_2 \cos\psi_2$, we obtain

$$\Delta n \cos\psi_1 = n_1 \Delta\psi \sin\psi_1 \quad . \tag{3.49}$$

Then, we take the following difference

$$n_2 \sin\psi_2 - n_1 \sin\psi_1 = \Delta n \sin\psi_1 + n_1 \Delta\psi \cos\psi_1 = \frac{\Delta n}{\sin\psi_1} \quad . \tag{3.50}$$

Here again, we have discarded second and higher order terms, and have substituted for $n_1\Delta\psi$ from (3.49).

Because $\sin\psi_1 = \Delta x_1/\Delta s_1$, and $\sin\psi_2 = \Delta x_2/\Delta s_2$ (Fig.3.28), (3.50) can be written as a difference equation

$$\frac{n_2\Delta x_2}{\Delta s_2} - \frac{n_1\Delta x_1}{\Delta s_1} = \frac{\Delta n}{\Delta x_1/\Delta s_1} \quad . \tag{3.51}$$

If the differences are now made infinitesimally small we obtain the following ray equation [3.1]

$$\frac{d}{ds}\left[n(s)\,\frac{dx(s)}{ds}\right] = \frac{dn(s)}{dx} \quad . \tag{3.52}$$

Here all variables have been expressed as functions of s, the path length along the ray.

This equation must be solved for the index profile of interest. As an example, let us consider the behaviour of a ray in a medium with a parabolic index profile, n(x), given by

$$n(x) = n_0[1 - \Delta(x/a)^2] \quad , \tag{3.53}$$

where $\Delta \ll 1$, and $|x| \leqq a$. We will also assume that the ray is incident almost axially or paraxially ($ds \approx dz$). Then, neglecting terms which are of higher order than x or its derivative, we find that the ray equation becomes

$$\frac{d^2x}{dz^2} + \frac{2\Delta}{a^2}\,x = 0 \quad . \tag{3.54}$$

This is the equation of the harmonic oscillator. The general solution is of the form

$$x(z) = x(0)\,\cos[(2\Delta)^{\frac{1}{2}}z/a] + x'(0)\,[a/(2\Delta)^{\frac{1}{2}}]\,\sin[(2\Delta)^{\frac{1}{2}}z/a] \quad , \tag{3.55}$$

where the prime denotes differentiation with respect to z.

We therefore obtain that the ray is confined to the inhomogeneous core and that its path is periodic in z. Figure 3.29 shows a set of such confined rays. It is apparent from (3.55) that all rays which are incident paraxially [$x'(0) \approx 0$] will focus periodically, as shown in Fig.3.29, and that the waveguide will behave like a cascade of lenses. Note that we made the assumption of paraxial incidence, so that we cannot allow the slope to vary much. For

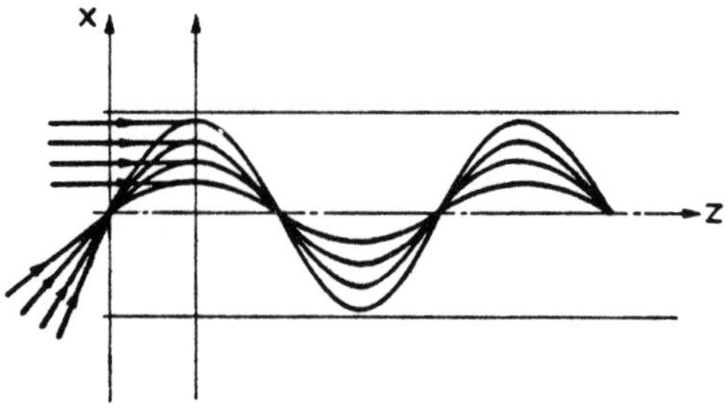

Fig. 3.29. Graphical illustration of (3.55) for axial and paraxial rays

the truly axial case, x'(0) would be zero, and the second term in (3.55) would disappear.

Let us now turn to the solution of Maxwell's equations

$$\underline{\nabla} \cdot \underline{D} = \sigma \quad \text{(a)} \qquad \underline{\nabla} \times \underline{E} = -\partial \underline{B}/\partial t \quad \text{(c)}$$
$$\underline{\nabla} \cdot \underline{B} = 0 \quad \text{(b)} \qquad \underline{\nabla} \times \underline{H} = \underline{J} + \partial \underline{D}/\partial t \quad \text{(d)} \tag{3.56}$$

where $\underline{E}$ and $\underline{H}$ are the electric and magnetic fields, respectively, while $\underline{D}$ and $\underline{B}$ are the corresponding displacement and induction. We assume source-free solutions, so that the charge density σ and the current density J are zero. Furthermore, in our inhomogeneous case, the dielectric constant is a function of position i.e. $\varepsilon(\underline{r}) = n^2(\underline{r})\varepsilon_0$. Then, assuming time harmonic solutions of the form $\exp(j\omega t)$, the curl equations reduce to

$$\underline{\nabla} \times \underline{E} = -j\omega\mu_0\underline{H} \quad ,$$
$$\underline{\nabla} \times \underline{H} = n^2(\underline{r})j\omega\varepsilon_0\underline{E} \quad . \tag{3.57}$$

By taking the curl of the first equation, multiplying the second by $(-j\omega\mu_0)$, and adding, we obtain the vector wave equation

$$\nabla^2\underline{E} + n^2(\underline{r})k_0^2\underline{E} = 0 \quad , \tag{3.58}$$

where $k_0^2 = \omega^2\varepsilon_0\mu_0$, and we have assumed that

$$\underline{\nabla} \cdot \underline{E} = -\underline{E} \cdot \underline{\nabla}\,[\ln n^2(\underline{r})] \approx 0 \quad . \tag{3.59}$$

A similar equation can be obtained for $\underline{H}$, provided that we fulfill the following condition

$$\underline{\nabla}[1/n^2(\underline{r})] \times (\underline{\nabla} \times \underline{H}) \ll \nabla^2\underline{H} \quad . \tag{3.60}$$

Now, in Cartesian coordinates, all components of the vector Laplacian in (3.58) separate and we obtain 3 scalar (Helmholtz) wave equations. We will be interested in the boundary condition that the tangential components of the field be continuous. For this purpose, we could use either the y component or the z component of the field. Consistent with the usage adopted by other authors, we will use the y component. Then, applying the separation of variable technique, we can split off the z dependence of the field by using the separation constant $-\beta^2$. This leads to a z dependence of the form $\exp(-j\beta z)$. Having split off the z dependence, the wave equation reduces to

$$\frac{d^2E_y}{dx^2} + [n^2(x)k_0^2 - \beta^2]E_y = 0 \quad . \tag{3.61}$$

(Note that the y variation is assumed to be zero.)

The differential equation (3.61) can be solved asymptotically for any index profile, using the WKB-method [3.31,32]. The idea is simply to obtain an expression for the direction of the local wavefront, as in geometrical optics. Thus, a phase function (called an eikonal in optics [3.1]) is introduced to express E_y as

$$E_y(x) = E_0 \exp[-jk_0 s(x)] \quad . \tag{3.62}$$

The complete solution then takes the form:

$$E_y(x,z,t) = E_0 \exp\{j[\omega t - \beta z - k_0 s(x)]\} \quad . \tag{3.63}$$

Before proceeding to the solution of the eikonal, let us digress a little and provide a geometrical interpretation for $s(x)$. For this purpose, let us express $s(x)$ as a Taylor series at points of continuity (say x_0)

$$s(x) = s(x_0) + (x - x_0)s'(x_0) + \ldots \quad . \tag{3.64}$$

If we ignore all but the two terms shown above, we can substitute back into (3.63) and interpret $k_0 s(x_0)$ to represent the local phase, and $k_0 s'(x_0)$ to be (approximately) the x component of the wave vector $\underline{k}$ (Fig.3.30).

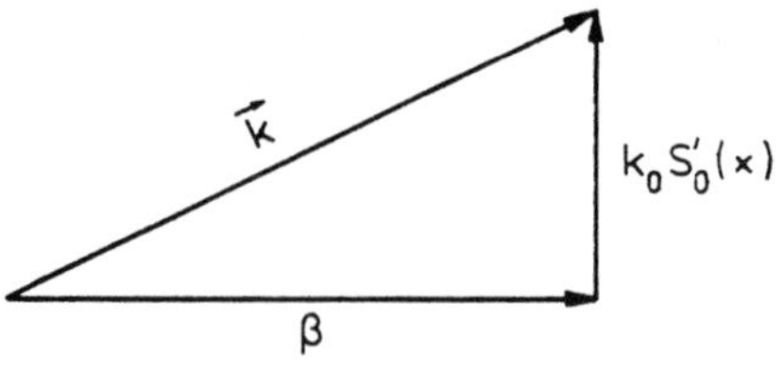

Fig. 3.30. Geometrical interpretation of the phase function $S_0'(x)$

Let us now return to the solution of (3.61) using (3.62). Direct substitution yields the nonlinear eikonal equation given below

$$-jk_0 s''(x) + [jk_0 s'(x)]^2 + [n^2(x)k_0^2 - \beta^2] = 0 \quad , \tag{3.65}$$

where the primes indicate differentiation with respect to x.

This equation can be recursively solved by using the perturbation method, with $k_0^{-1} = \lambda_0/(2\pi)$ as the small parameter. We first expand s(x) as follows

$$s(x) = s_0(x) + s_1(x)/k_0 + s_2(x)/k_0^2 + \dots \quad . \tag{3.66}$$

By substituting this expansion into (3.65), a bulky expression is obtained in terms of decreasing powers of k_0. The expression should vanish identically for any k_0, so that each coefficient of k_0^{-n} must vanish for n = -2, -1, 0, ... Hence,

$$\left.\begin{array}{ll} -k_0^2 s_0'^2 + k_0^2 n^2(x) - \beta^2 & = 0 \\ -js_0'' - 2s_0's_1' & = 0 \\ -js_1'' - s_1'^2 - 2s_0's_2' & = 0 \\ \vdots & \vdots \\ & = 0 \end{array}\right\} \tag{3.67}$$

We can now observe that the first equation of this set can be solved for s_0, the second for s_1, the third for s_2, etc. Usually, all but s_0 and s_1 are neglected, and we simply obtain

$$\left.\begin{aligned} & s_0(x) = \pm \int^{x} F(x)dx \quad , \\ \text{and} & \\ & s_1(x) = \frac{-j}{4} \ln[F^2(x)] \quad , \\ \text{with} & \\ & F(x) = \left[n^2(x) - \beta^2/k_0^2\right]^{\frac{1}{2}} \quad . \end{aligned}\right\} \tag{3.68}$$

Replacing $s(x)$ in (3.62) by s_0 and s_1 from (3.68), yields the following expression for the field

$$E_y(x) = \frac{E_0}{\sqrt{F(x)}} \exp\left[\pm jk_0 \int^{x} F(x)dx\right] \quad . \tag{3.69}$$

We note that (3.69) actually includes two solutions: one corresponding to the + sign and one to the - sign. The nature of the WKB solution can, as before, be interpreted by examining $s_0'(x)$ which determines the x component of $\underline{k}$ (Fig.3.31)

$$s_0'(x) = F(x) = \sqrt{n^2(x) - \beta^2/k_0^2} \quad . \tag{3.70}$$

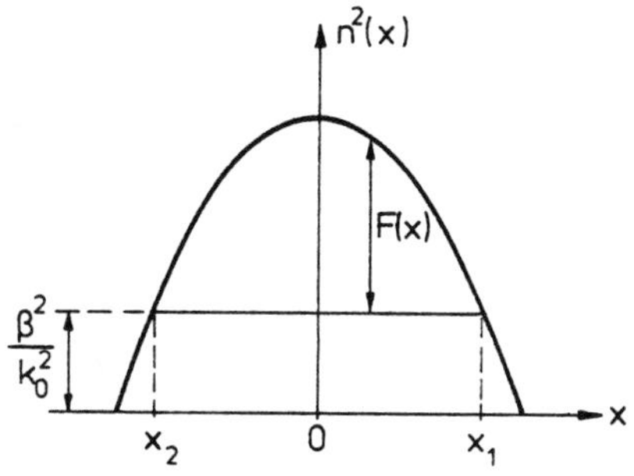

Fig. 3.31. Graphical interpretation of (3.70) showing propagation and radiation regimes

Thus, $s_0'(x)$ is zero for $n^2(x) = \beta^2/k_0^2$. When $n^2(x) > \beta_1^2/k_0^2$, $s_0'(x)$ is real, and we have periodic solutions as in (3.71) below. In contrast to this, when $n^2(x) < \beta^2/k_0^2$, $s_0'(x)$ takes the form $j|F(x)|$, and we obtain aperiodic solutions as in (3.72):

$$E_y(x) = \frac{E_1}{\sqrt{F}} \exp\left(jk_0 \int_{x}^{x_1} Fdu\right) + \frac{E_2}{\sqrt{F}} \exp\left(-jk_0 \int_{x}^{x_1} Fdu\right) \tag{3.71}$$

$$E_y(x) = \frac{E_3}{\sqrt{j|F|}} \exp\left(k_0 \int_{x_1}^{x} |F| du\right) + \frac{E_4}{\sqrt{j|F|}} \exp\left(-k_0 \int_{x_1}^{x} |F| du\right) \tag{3.72}$$

Note that each of the above two equations is a linear combination of the two solutions represented by (3.69). We know from previous sections that the field beyond the turning point should be decaying. Thus, we take $E_3 = 0$ in (3.72). Furthermore, we know, both from the foregoing and from our previous ray path analysis, that our guide has another turning point besides x_1 (say x_2) which is negative. In this case, we choose the other solution, and take $E_4 = 0$. We therefore obtain

$$E_y(x) = \frac{E_4}{\sqrt{j|F|}} \exp\left(-k_0 \int_{x_1}^{x} |F| du\right) , \qquad \text{for } x > x_1 , \tag{3.73}$$

and

$$E_y(x) = \frac{E_3}{\sqrt{j|F|}} \exp\left(-k_0 \int_{x}^{x_2} |F| du\right) , \qquad \text{for } x < x_2 . \tag{3.74}$$

We are now ready to satisfy boundary conditions, which require that (3.73) and (3.74) be identical to (3.71) at the turning points x_1 and x_2. Unfortunately, we have a problem! When $x = x_1$ or x_2, $F(x) = 0$, and all our solutions become infinite. We know that physically this cannot be, and conclude that the WKB approximation loses validity around the turning points. So, what are we to do now?

We return to (3.61) and try to find a better appproximation for the fields around the turning points. We note that the difference $n^2(x)k_0^2 - \beta^2$ becomes small near the turning points, so that we should be able to expand this expression in a Taylor series around x_1. Thus, neglecting second and higher order terms, we can write

$$n^2(x)k_0^2 - \beta^2 \approx -k_0^2 N_1(x - x_1) , \tag{3.75}$$

where $N_1 = -2n\, dn/dx$ at $x = x_1$. Then, using a change of variable $w = k_0^{2/3} N_1^{1/3}(x - x_1)$, (3.61) becomes

$$\frac{d^2E_y}{dw^2} - wE_y = 0 . \tag{3.76}$$

This is the Airy differential equation with solutions of the form Ai(w) and Bi(w). The first of these suits our purposes, as can be seen from Fig.3.32. We notice that Ai(w) decays beyond $w = 0$ (or $x = x_1$), and oscillates for negative $w(x < x_1)$. Asymptotic solutions to (3.76) can also be found [3.33] for $w > 1$ and $w < -5$, and in these cases they take the following forms (for positive and negative arguments)

$$E_y(w) = \frac{E}{2\pi^{1/2}w^{1/4}} \exp\left(-\frac{2}{3}w^{3/2}\right) , \tag{3.77}$$

and

$$E_y(-w) = \frac{E}{\pi^{1/2}w^{1/4}} \sin\left(\frac{2}{3}w^{3/2} + \pi/4\right) . \tag{3.78}$$

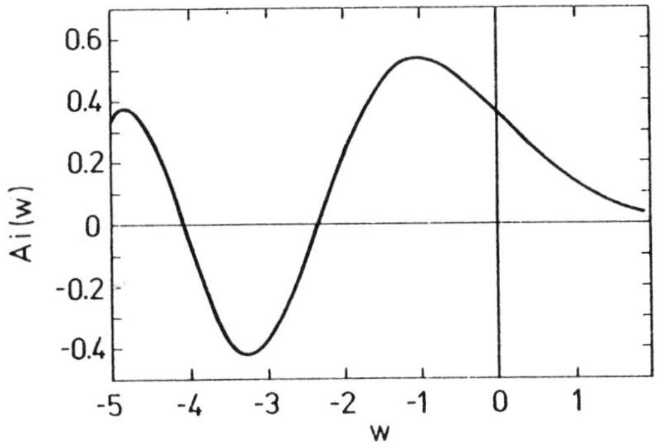

Fig. 3.32. Plot of the Airy function Ai(w)

If we compare these equations with the WKB solutions (3.71,73, and 74), we notice that they would be rather similar for suitable arguments and constants. Let us replace the arguments in (3.77 and 78) by the following expressions

$$\begin{aligned} \frac{2}{3}w^{3/2} &= k_0 \int_{x_1}^{x} |F(u)|du \qquad x > x_1 \\ &= k_0 \int_{x}^{x_1} F(u)du \qquad x < x_1 \quad . \end{aligned} \tag{3.79}$$

Then, differentiation of (3.79) and substitution into (3.77 and 78) yields the following results

$$E_y(x) = \frac{E_5 N_1^{1/6} k_0^{1/3}}{2\sqrt{\pi|F(x)|}} \exp\left[-k_0 \int_{x_1}^{x} |F(u)|du\right] \quad \text{for } x > x_1 \quad , \tag{3.80}$$

and

$$E_y(x) = \frac{E_5 N_1^{1/6} k_0^{1/3}}{\sqrt{\pi F(x)}} \sin\left[k_0 \int_x^{x_1} F(u)du + \pi/4\right] \quad \text{for } x < x_1 \quad . \qquad (3.81)$$

Similarly, the solutions around the other turning point x_2 take the form:

$$E_y(x) = \frac{E_6 N_2^{1/6} k_0^{1/3}}{2\sqrt{\pi}\, F(x)} \exp\left[-k_0 \int_x^{x_2} |F(u)|du\right] \quad \text{for } x < x_2 \quad , \qquad (3.82)$$

and

$$E_y(x) = \frac{E_6 N_2^{1/6} k_0^{1/3}}{\sqrt{\pi F(x)}} \sin\left[k_0 \int_{x_2}^{x} F(u)du + \pi/4\right] \quad \text{for } x > x_2 \quad . \qquad (3.83)$$

In putting the solutions in this form, we have assumed that Airy solutions provide a sufficiently accurate picture even far from the turning points, and that the constants of the WKB solutions can be suitably tailored to match the Airy solutions. Referring now to (3.81 and 83), we note that these two equations must be identical for a unique field description. This is true if and only if $E_5 N_1^{1/6} = E_6 N_2^{1/6}$ and the sum of the arguments of sines is an odd multiple of π, or $E_5 N_1^{1/6} = -E_6 N_2^{1/6}$ and the sum of the arguments is an even multiple of π. Thus, for $\mu = 0, 1, 2, \ldots$, and re-introducing the original definition of $F(u)$, we have

$$\int_{x_2}^{x_1} \left[k_0^2 n^2(x) - \beta^2\right]^{\frac{1}{2}} dx = (\mu - 1/2)\pi \quad . \qquad (3.84)$$

This is the eigenvalue equation of graded index waveguides. We note that each integer μ defines some mode with a particular value of the axial propagation constant β. Thus, once again we have the situation that only a certain number of modes can be supported by the waveguide.

3.5 Step-Index Fibers

In this section we will examine the simplest of all cylindrical waveguides commonly known as the step-index fiber. This form of dielectric waveguide was studied as long ago as 1910 by HONDROS and DEBYE [3.34], and more recently by KAPANY and BURKE [3.35], MARCUSE [3.28], GLOGE [3.36], and UNGER [3.29], amongst others. The basic structure, shown in Fig.3.33, consists of a circular

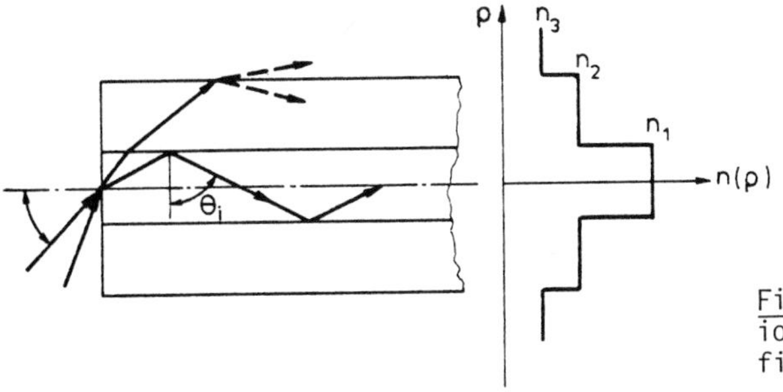

Fig. 3.33. The path of meridional rays in a step-index fiber

core of refractive index n_1, and a cladding region of index n_2, with n_2 less than n_1. The circular interface thus obtained provides the basis for propagation, but for strength and protection the cladding is usually covered by a third absorbing medium of refractive index n_3. However, the cladding region must be made sufficiently thick in order to prevent this third medium from absorbing energy meant for propagation. With this justification, we will assume that the cladding extends to infinity and that we have only one interface. Then, if a meridional ray strikes the interface at an angle θ_i which is greater than the critical angle θ_{ic}, it is totally internally reflected (Fig. 3.33), and propagates along the fiber in a zig-zag path. If such a ray is incident on the end face of the fiber at an angle θ, then using geometrical optics it is easy to show that propagation takes place for

$$\sin\theta \leqq (n_1^2 - n_2^2)^{\frac{1}{2}} \quad . \tag{3.85}$$

The equality represents the maximum acceptance angle (say θ_0) of the fiber. In practice, $\sin\theta_0$, known as the numerical aperture of the fiber, is the parameter used for describing the maximum acceptance angle.

Let us now apply some of the results of the previous sections to qualitatively see the behaviour of plane waves and bounded beams at our circular interface. Firstly, we note that the meridional case is somewhat similar to the planar waveguide situation of Section 3.3. Thus, for a plane wave, we obtain propagation for $\theta_i \geqq \theta_{ic}$ for certain discrete angles of incidence which satisfy the boundary conditions. Similarly, for $\theta_i < \theta_{ic}$, the wave is refracted out of the core. Furthermore, for propagating waves, we know that energy is not just confined to the core but extends exponentially into the cladding region. Thus, a sufficiently thick cladding is required to ensure that the tail of the exponential would be approximately zero at the outer edge of the cladding. If we replace the plane wave by a bounded beam, we effectively have a set of plane waves incident at a variety of angles - the narrower the beam, the larger the variety. For incidence sufficiently close to critical, some of the plane waves would then be refracted, leading to loss of energy from the beam. In this meridional case, the first reflection suffered by the beam acts as a filtration process and removes most of the unsuitable components of the bounded beam.

However, unlike in planar waveguides, in a fiber we may also launch skew waves, as shown in Fig.3.34. A plane wave incident in this manner "sees" the curvature of the fiber since every plane of incidence is at an angle to the axis. Consequently, the optical interface is effectively an ellipse and the results of Sect.3.2 can be applied. It is then obvious that every point on the plane wavefront will see a different part of the ellipse, and that the propagation vector associated with each point will see a different angle of incidence. We now have the situation in which a part of the plane wave will be totally internally reflected, but other parts will be incident at angles less than or equal to critical. Part of this energy will tunnel through a "forbidden" region and reappear as a propagating wave in the cladding, while a part will be simply refracted.

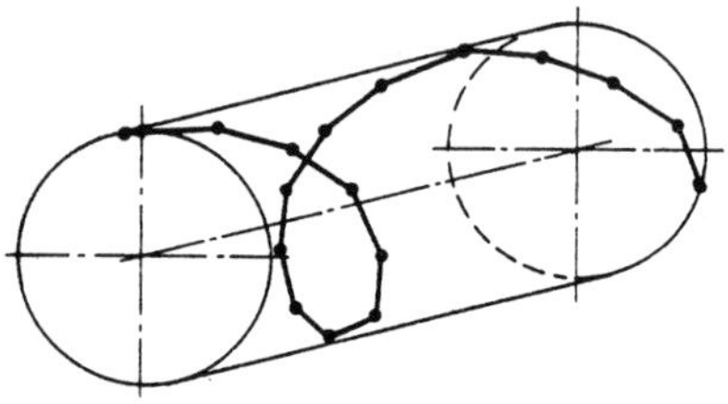

Fig. 3.34. Skew rays in a step-index fiber

We should now note that the plane wave situation described above can never exist in a fiber, because the input aperture of the fiber will immediately limit any incident "plane wave". After the first incidence, the wave will suffer further distortion due to reflection, refraction, and tunnelling, and will assume some strange plane wave spectrum. Hence, the behaviour of such a wave must be examined on the basis of the behaviour of individual plane wave components. Similarly, a bounded beam incident in a skew manner in the fiber, will also rather quickly lose its shape. However, if the beam is narrow, it would not be so sensitive to curvature, and its initial changes would be mostly due to refractive leakage. We should also note that, due to the elliptical interface encountered by skew waves, the curvature seen by such waves is different at each incidence. We would therefore not expect reflection, refraction, and tunnelling to have the same relative influence at every incidence. In fact, we would expect that refractive leakage would end within a relatively short distance, since worst case curvatures would have been encountered by then. However, energy leakage by tunnelling could continue for long distances. In any case, we could conclude that a skew wave can form both a propagating and a leaky mode of the fiber, and that a leaky mode must eventually lose all its energy by tunnelling. Energy leakage by refraction must not be confused with the energy leakage by tunnelling which occurs in a leaky mode.

At this point it should be noted that in contrast to planar guides, fibers are generally required to carry power over substantially longer distances: consider communications networks! As a result, the modulation characteristics of the fiber become of prime interest. These characteristics are primarily affected by material and mode dispersions, which cause any input pulse of light to spread during propagation. In our treatment, we will ignore material dispersion, as it is essentially a material characteristic. Furthermore, since lasers are often used as sources and have narrow spectral widths, material dispersion becomes a second order effect, particularly in multimode step index fibers. As such, we will concentrate on mode dispersion.

We already know from our previous considerations (Sect.3.3), that a plane wave can only propagate at discrete angles in a waveguide, even though incident at angles greater than the critical. Each suitable angle represents a guided "mode" of the waveguide, and we see that each mode must propagate along a different path. In a fiber, we would expect a number of meridional modes, as well as many skew modes, which together would represent all the guided modes of the fiber. If we would excite all the fiber modes using a Dirac impulse of light, then the output from a sufficiently long fiber would be a train of impulses spread out in time. Furthermore, if a large number of modes would exist, then the output would tend to be continuous, particularly since material dispersion would in any case broaden each impulse. To determine the impulse response, we have to know the effective waveguide velocity of each of the allowed modes. This velocity is given by the well known group velocity formula found in most textbooks of optics, e.g. [3.1], and can be simply applied to our case of a homogeneous core (step index) to yield the propagation time

$$\tau_{\mu\nu} = \frac{L}{c} \frac{\partial \beta_{\mu\nu}}{\partial k_0} \quad . \tag{3.86}$$

Here, β is the axial propagation constant of a given mode, $k_0 = 2\pi/\lambda$, L is the length in question, and c is the velocity of light in vacuo. The material has also been assumed to be non-dispersive. Moreover, we have anticipated that each mode can be characterized by integers μ and ν (to be discussed later). To simplify calculations, we will assume that each mode carries the same amount of power, in which case the impulse response can be simply written as

$$h(t) = \sum_{\mu\nu} \delta(t - \tau_{\mu\nu}) \quad , \tag{3.87}$$

where $\delta(t - \tau_{\mu\nu})$ is a Dirac impulse located at $\tau_{\mu\nu}$.

It is also more convenient, from the viewpoint of numerical calculations, to assume an input pulse shape, say S(t), whose width is small compared to the impulse response. Then h(t) can be approximated as follows

$$h(t) \simeq S(t) * \sum_{\mu\nu} \delta(t - \tau_{\mu\nu}) = \sum_{\mu\nu} S(t - \tau_{\mu\nu}) \quad , \tag{3.88}$$

where the symbol (*) represents convolution.

Having described the basis of the impulse response calculations, it becomes clear that we must calculate the allowed propagation constants, their relationship to k_0, and hence their propagation delays $\tau_{\mu\nu}$. To achieve this end, we must solve Maxwell's equations for the fiber, and obtain the eigenvalue equation. (For an in-depth study of the fields in a step-index fiber, we refer the reader to the excellent and detailed book by UNGER [3.29]). We therefore start with Maxwell's equations (3.56), and as in Sect.3.4 obtain the time independent vector wave equation, but now for a constant refractive index of the core. The equation then becomes

$$\underline{\nabla} \cdot \underline{\nabla}\, \underline{E} + k_n^2 \underline{E} = 0 \quad , \tag{3.89}$$

where $k_n = 2\pi n/\lambda$.

In cylindrical coordinates, with the fiber along the z axis, the z component separates [3.37], and we obtain the scalar Helmholtz equation:

$$\underline{\nabla}^2 E_z + k_n^2 E_z = 0 \quad . \tag{3.90}$$

Casting this into cylindrical coordinates gives

$$\frac{1}{\rho}\frac{\partial}{\partial\rho}\left(\rho\frac{\partial E_z}{\partial\rho}\right) + \frac{1}{\rho^2}\frac{\partial^2 E_z}{\partial\phi^2} + \frac{\partial^2 E_z}{\partial z^2} + k_n^2 E_z = 0 \quad . \tag{3.91}$$

This partial differential equation can be reduced to a set of ordinary differential equations using the "separation of variables" technique. For this purpose we try solutions of the form

$$E_z(\rho,\phi,z) = F(\rho)\Phi(\phi)Z(z) \quad , \tag{3.92}$$

and split off the z dependence using the separation constant $-\beta^2$, and the ϕ dependence using the separation constant $-\nu^2$. This gives

$$\left.\begin{aligned} &\frac{1}{Z}\frac{d^2 Z}{dz^2} = -\beta^2 \quad , \\ &\frac{1}{\Phi}\frac{d^2\Phi}{d\phi^2} = -\nu^2 \quad , \\ &\text{and} \\ &\rho^2\frac{d^2F}{d\rho^2} + \rho\frac{dF}{d\rho} + \left[\rho^2(k_n^2 - \beta^2) - \nu^2\right]F = 0 \quad . \end{aligned}\right\} \tag{3.93}$$

The solution $\exp(-j\beta z)$ for the z dependence represents a wave propagating in the z direction. For the ϕ dependence, we select a solution of the form $\cos\nu\phi$ for the E field, and reserve $\sin\nu\phi$ for the H field. This ensures orthogonality of the E and H fields. Choice of $\cos\nu\phi$ for the E field implies that the coordinate axes must be suitably aligned with respect to the fiber. In any case, the important feature to notice is that, ϕ being the azimuthal coordinate, we must have

$$\cos\nu\phi = \cos\nu(\phi + 2\pi) \quad , \tag{3.94}$$

or that ν must be an integer. Note that we have now obtained one of the integers used to describe the propagation constant in (3.86).

The third equation in (3.93) must be solved separately for the core and cladding regions. If the refractive index in the core region is n_1, then k_n must be replaced by $2\pi n_1/\lambda$. A change of variable then yields the Bessel differential equation for the core [3.33]

$$w^2 \frac{d^2F}{dw^2} + w \frac{dF}{dw} + (w^2 - \nu^2) F = 0 \quad , \tag{3.95}$$

where $w = \rho\gamma_1$, with $\gamma_1^2 = (2\pi n_1/\lambda)^2 - \beta^2$.

For proper behaviour near the core axis, we choose solutions $J_\nu(\gamma_1\rho)$, since this function is bounded for $\gamma_1\rho \to 0$.

In the cladding region, β^2 exceeds $(2\pi n_2/\lambda)^2$, so that the term γ_2 becomes imaginary. We replace γ_1 in the above by $j\gamma_2$, and use a change of variable $w_1 = \gamma_2\rho$ to obtain the modified Bessel equation:

$$w_1^2 \frac{d^2F}{dw_1^2} + w \frac{dF}{dw_1} - (w_1^2 - \nu^2) F = 0 \quad . \tag{3.96}$$

The proper decaying behaviour in the cladding is yielded by the modified Hankel functions $K_\nu(\gamma_2\rho)$, since $K_\nu(\gamma_2\rho) \to 0$ for $w_1 \to \infty$.

Inserting these solutions back into (3.92) and bearing in mind the original assumption of a temporally harmonic wave, the fields in the core and cladding become

$$\left.\begin{aligned} E_{z_1} &= A\, J_\nu(\gamma_1\rho) \cdot \cos\nu\phi \cdot \exp[j(\omega t - \beta z)] \quad , \\ H_{z_1} &= B\, J_\nu(\gamma_1\rho) \cdot \sin\nu\phi \cdot \exp[j(\omega t - \beta z)] \quad , \\ E_{z_2} &= C\, K_\nu(\gamma_2\rho) \cdot \cos\nu\phi \cdot \exp[j(\omega t - \beta z)] \quad , \\ H_{z_2} &= D\, K_\nu(\gamma_2\rho) \cdot \sin\nu\phi \cdot \exp[j(\omega t - \beta z)] \quad . \end{aligned}\right\} \tag{3.97}$$

The boundary conditions to be satisfied are that the tangential field components must be continuous across the boundary. We require $E_{z_1} = E_{z_2}$, $H_{z_1} = H_{z_2}$, $E_{\phi_1} = E_{\phi_2}$, and $H_{\phi_1} = H_{\phi_2}$. We see that the first two conditions can be satisfied by using (3.97), but we must also solve for the ϕ field components.

Fortunately, both the ϕ and ρ components can be obtained from the z components by using Maxwell's curl equations and assuming a temporally harmonic dependence, $\exp(j\omega t)$. The time independent curl equations thus obtained are first cast into cylindrical coordinates. Then, the ϕ and ρ components on the left hand and right hand side of the curl equations are equated, to give a set of four simultaneous equations as follows

$$
\left.
\begin{aligned}
-j\omega\mu H_\rho &= \frac{1}{\rho}\frac{\partial E_z}{\partial \phi} - \frac{\partial E_\phi}{\partial z} \, , \\
-j\omega\mu H_\phi &= \frac{\partial E_\rho}{\partial z} - \frac{\partial E_z}{\partial \rho} \, , \\
j\omega\varepsilon E_\rho &= \frac{1}{\rho}\frac{\partial H_z}{\partial \phi} - \frac{\partial H_\phi}{\partial z} \, , \\
j\omega\varepsilon E_\phi &= \frac{\partial H_\rho}{\partial z} - \frac{\partial H_z}{\partial \rho} \, .
\end{aligned}
\right\} \qquad (3.98)
$$

We have four unknowns and four equations, which can be solved to give the following results

$$
\left.
\begin{aligned}
E_\phi &= -\frac{j}{\gamma^2}\left(\frac{\beta}{\rho}\frac{\partial E_z}{\partial \phi} - \omega\mu\frac{\partial H_z}{\partial \rho}\right) \, , \\
E_\rho &= -\frac{j}{\gamma^2}\left(\frac{\mu\omega}{\rho}\frac{\partial H_z}{\partial \phi} + \beta\frac{\partial E_z}{\partial \rho}\right) \, , \\
H_\phi &= -\frac{j}{\gamma^2}\left(\omega\varepsilon\frac{\partial E_z}{\partial \rho} + \frac{\beta}{\rho}\frac{\partial H_z}{\partial \phi}\right) \, , \\
H_\rho &= -\frac{j}{\gamma^2}\left(\beta\frac{\partial H_z}{\partial \rho} - \frac{\omega\varepsilon}{\rho}\frac{\partial E_z}{\partial \phi}\right) \, .
\end{aligned}
\right\} \qquad (3.99)
$$

Consequently, when using cylindrical coordinates, the ϕ and ρ components of the field can always be obtained from the z components (both for the core and for the cladding) irrespective of the refractive index profile. In any case, we can now substitute the fields from (3.97) into (3.99) and obtain the ϕ components for the core and cladding. We then satisfy boundary conditions by equating the z and ϕ components of the electric and magnetic field. This results in a set of four homogeneous equations in the field constants A, B, C, and D. We can use the determinant method for solving this set of simultaneous equations which are homogeneous and therefore have non-zero solutions (for A, B, C, and D) only when the determinant of their coefficients is zero. This determinant is given below

$$\begin{vmatrix} J_\nu & 0 & -K_\nu & 0 \\ 0 & J_\nu & 0 & -K_\nu \\ \frac{\beta\nu}{a\gamma_1^2} J_\nu & \frac{\omega\mu_1}{\gamma_1} J'_\nu & \frac{\beta\nu}{a\gamma_2^2} K_\nu & \frac{\omega\mu_2}{\gamma_2} K'_\nu \\ \frac{\omega\varepsilon_1}{\gamma_1} J'_\nu & \frac{\beta\nu}{a\gamma_1^2} J_\nu & \frac{\omega\varepsilon_2}{\gamma_2} K'_\nu & \frac{\beta\nu}{a\gamma_2^2} K \end{vmatrix} = 0 \quad , \tag{3.100}$$

where a is the radius of the core, and J'_ν and K'_ν are derivatives with respect to the radial coordinate ρ, evaluated at $\rho = a$. J_ν, J'_ν are understood to have arguments $(\gamma_1 a)$ while K_ν, K'_ν to have arguments $(\gamma_2 a)$.

Expansion of the above determinant leads to the often misquoted and/or misprinted eigenvalue equation

$$\frac{\beta^2\nu^2}{a^2}\left[\frac{1}{\gamma_1^2}+\frac{1}{\gamma_2^2}\right]^2 = \left[\frac{J'_\nu(\gamma_1 a)}{\gamma_1 J_\nu(\gamma_1 a)} + \frac{K'_\nu(\gamma_2 a)}{\gamma_2 K_\nu(\gamma_2 a)}\right]\left[\frac{k_1^2}{\gamma_1}\frac{J'_\nu(\gamma_1 a)}{J_\nu(\gamma_1 a)} + \frac{k_2^2}{\gamma_2}\frac{K'_\nu(\gamma_2 a)}{K_\nu(\gamma_2 a)}\right] \quad . \tag{3.101}$$

We can now reach several important conclusions:

1) Although the eigenvalue equation is unwieldy, it contains only one unknown, β.

2) Due to the oscillatory character of Bessel functions, for every value of the integer ν, there exists μ allowed solutions for β, since allowed values of β must lie in the range (see Sect.3.3)

$$\frac{2\pi n_2}{\lambda} \leqq \beta \leqq \frac{2\pi n_1}{\lambda} \quad .$$

Alternatively, we can say that each allowed β is characterized by integers μ and ν. This explains our use of the symbol $\beta_{\mu\nu}$ in (3.86).

3) The field constants A, B, C, and D in (3.97) can only be determined if we specify the value of one of them. Thus, for example, we can determine B, C, and D in terms of A, which is then called the excitation coefficient. The relative values of these coefficients are different for different modes, so that for some modes the longitudinal electric field dominates the longitudinal magnetic field, while for other modes the situation may be vice versa. Yet other modes ($\nu = 0$) may have zero longitudinal electric or magnetic fields.

4) By convention, $HE_{\mu\nu}$ modes are those whose longitudinal electric field dominates their longitudinal magnetic field. In the opposite case, the modes are classified as $EH_{\mu\nu}$. If the longitudinal electric field is zero, we have $TE_{0\mu}$ modes, and if the longitudinal magnetic field is zero, the modes are tagged $TM_{0\mu}$.

5) Finally, we recognize that the eigenvalue (3.101) can only be solved numerically.

Returning now to the eigenvalue equation, we adopt UNGER's notation [3.29] and cast (3.101) into the following form:

$$\frac{N^2\nu^2}{\left(u^2B\right)^2} = \left(Y_\nu + X_\nu\right)\left(n_1^2Y_\nu + n_2^2X_\nu\right) \quad , \tag{3.102}$$

where $u = \gamma_1 a$, $v = \gamma_2 a$, $N = \beta/k_0$, $B = (N^2 - n_2^2)/(n_1^2 - n_2^2)$, $X_\nu = K_\nu'(v)/[vK_\nu(v)]$, and $Y_\nu = J_\nu'(u)/[uJ_\nu(u)]$. As written above, the eigenvalue equation is quadratic in Y_ν (or X_ν) [3.35]. We can solve for Y_ν, for example, and if we assume a weakly guiding fiber ($n_1 \approx n_2$), the solution reduces to

$$Y_\nu + X_\nu \approx \pm \frac{\nu}{u^2B} \quad , \quad (n_1 \approx n_2) \quad . \tag{3.103}$$

Using standard identities [3.33], we can write X_ν and Y_ν in the form

$$Y_\nu = \frac{J_\nu'(u)}{uJ_\nu(u)} = \pm \frac{J_{\nu\mp1}(u)}{uJ_\nu(u)} \mp \frac{\nu}{u^2} \quad , \tag{3.104}$$

$$X_\nu = \frac{K_\nu'(v)}{vK_\nu(v)} = - \frac{K_{\nu\mp1}(v)}{vK_\nu(v)} \mp \frac{\nu}{v^2} \quad . \tag{3.105}$$

Then, using the upper sign with the negative root, and the lower with the positive, we obtain the following approximate form of the eigenvalue equation

$$\frac{uJ_\nu(u)}{J_{\nu\mp1}(u)} \approx \pm \frac{vK_\nu(v)}{K_{\nu\mp1}(v)} \quad , \quad (n_1 \approx n_2) \quad . \tag{3.106}$$

Comparison of (3.101) with the last result, clearly shows the reduction in complexity which results from assuming nearly equal refractive indices. We now note some of the features (due to this eigenvalue equation) of the so-called weakly guiding fiber.

1) The eigenvalue equation (3.106) is in fact two equations in one. The plus sign on the right hand side goes with the (ν-1) index, and gives rise to one equation, while the minus sign on the right hand side goes with the (ν + 1) index, and gives the second equation. This leads to HE modes for the plus sign and EH modes for the minus sign.

2) If we apply standard recurrence relations (see e.g. [3.33]) to the eigenvalue equation for HE modes, we obtain

$$\frac{uJ_{\ell}}{J_{\ell+1}} + \frac{vK_{\ell}}{K_{\ell+1}} = 0 \quad , \tag{3.107}$$

where $\ell = \nu - 2$. This is of the same form as the equation for EH modes except for the index difference. We can thus conclude that $HE_{\nu,\mu}$ modes are degenerate with $EH_{\nu-2,\mu}$ modes i.e. their β values are equal. Furthermore, $HE_{1,\mu}$ modes have special significance since they can only be degenerate with the nonexistent $EH_{-1,\mu}$ modes. Thus, if we would solve for allowed β values only using the equation for EH modes, we would then lose all $HE_{1,\mu}$ modes. On the other hand, if we only use the equation for HE modes, we must then account for the fact that $HE_{1,\mu}$ modes are not degenerate. Another feature of the degeneracy between HE and EH modes is that it allows us to make a linear combination of field solutions. Upon doing this, we find that the ρ and ϕ field components are such as to yield only E_x and H_y, or E_y and H_x components, depending on the axis definition. Furthermore, for the weakly guiding case, the E_z component is also very small, so that it becomes reasonable to talk about linearly polarized modes. This is the origin of the term $LP_{\ell p}$ modes.

At this point, we digress somewhat and ask the question: how many modes does a step-index fiber support and under what conditions does it carry only a single mode? To answer this question, we need the so called "V" parameter, given by

$$V^2 = u^2 + v^2 = (n_1^2 - n_2^2)(2\pi a/\lambda)^2 \quad . \tag{3.108}$$

Then, as will be seen in Section 3.6, the number of modes is approximately given by

$$M = 0.5\ V^2 \quad . \tag{3.109}$$

The condition under which only the HE_{11} mode propagates, can also be given in terms of V (see e.g. [3.29]) and represents the limit of single mode operation:

$$V < 2.405 \quad . \tag{3.110}$$

For example, a fiber with $a = 20\ \mu m$, $n_1 = 1.54$, and an index difference of 0.5%, would be expected to support approximately 258 modes at 850 nm. If the indices and wavelength would be kept constant, a diameter of 4 μm would ensure single mode operation. Alternatively, to obtain a larger diameter, the index difference would have to be reduced.

Returning now to the calculation of the impulse response, we note that allowed β values can be found either from the exact eigenvalue equation (3.101) or from its approximation (3.106). From (3.86), we see that we must also calculate $d\beta_{\nu\mu}/dk_0$ to obtain $\tau_{\mu\nu}$. A simple approach is to estimate the derivative from the incremental change $\Delta\beta/\Delta k_0$ at each β of interest. We can then use (3.88) to estimate the impulse response.

Figure 3.35 shows the separate response of HE and EH modes [calculated from (3.106)] to an input Gaussian pulse with a width of 3 ns [3.38]. The figure also shows the impulse response obtained from the exact eigenvalue equation (3.101). The fiber has been assumed to have the following parameters: length 1 km, core radius 20 μm, core index 1.54, core-cladding index difference 1.5%, and zero material dispersion. The source wavelength has been assumed to be 850 nm. We see that the impulse responses for HE and EH modes (Curves 2 and 3) are almost the same and that the peaks arrive at approximately similar times. Thus, our previous conclusions about the degeneracy between HE and EH modes would seem to hold even for an index difference of 1.5%. Note, however, that the exact solution (Curve 3) gives an impulse response which is considerably smoother, with only one peak. The leading edges of all three curves are almost coincident, but the trailing edge of the exact response arrives significantly earlier and predicts a narrower impulse response. To see the possible improvement due to a smaller index difference, we present the curves of Fig.3.36 for the same parameters as before but with an index difference of 0.5% [3.38]. The impulse response using the so called WKB approximation (Sect.3.6) is also presented (Curve 3). We see that although the exact and "weakly guided" solutions are now closer to each other, they are still noticeably different. Again, the leading edges are almost coincident but the trailing edges are separated much more. Even stronger deviations can be seen for the case of the WKB solution which seems to predict a significantly broader impulse response and a smoother mode distribution. (Note that the WKB impulse response has been obtained by first analytically solving the

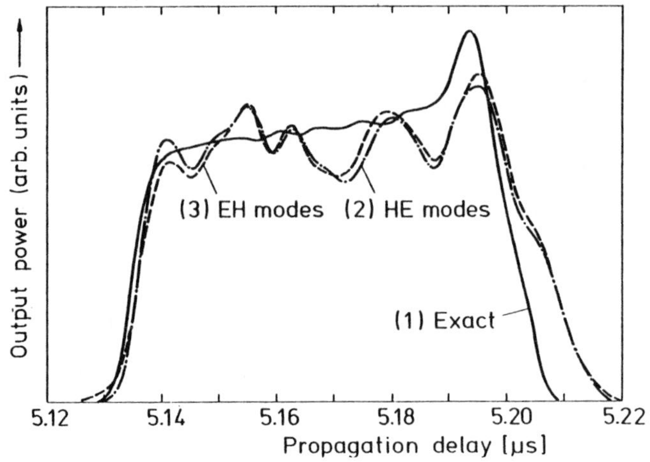

Fig. 3.35

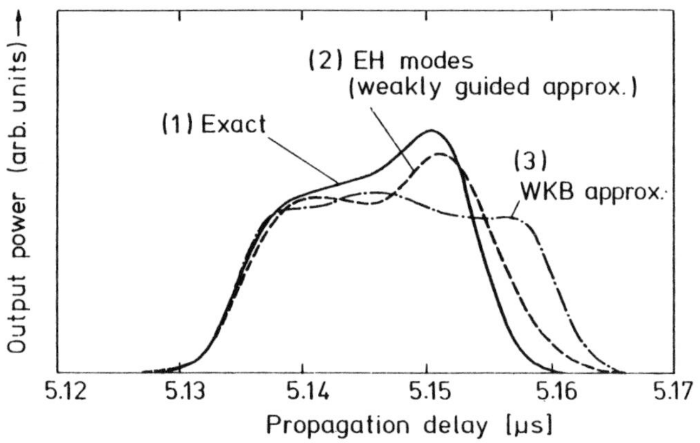

Fig. 3.36

Fig. 3.35. Impulse response of a step-index fiber with a 1.5% index difference and other parameters as in the text: (1) Exact solution, (2) HE modes only, and (3) EH modes only

Fig. 3.36. Impulse response of a step-index fiber with a 0.5% index difference, and the other parameters as in Fig.3.35: (1) Exact solution, (2) EH modes only, and (3) the WKB approximation

WKB eigenvalue integral as in Sect.3.6). It becomes clear from these curves that, although the various approximations yield roughly the correct response, they must be used with a degree of caution.

We conclude this section by discussing the implication of two of our assumptions which can significantly affect the impulse response. We have explicitly assumed that *all* modes are excited and excited equally; also, we have implicitly assumed that there is no intermodal power exchange or mode coupling. From a practical view-point, neither of these conditions can be completely fulfilled because (I) fiber imperfections, such as microbends, inevitably cause mode coupling, and (II) one may deliberately wish to have selective excitation, for example, to improve launching efficiency. It is then clear that a fiber which has been selectively excited, but has little mode coupling, will exhibit an impulse response significantly different from the calculated one. Even for a long fiber, in which all modes would be eventually excited, we would expect deviations from the calculated response, because different sections of the fiber would behave differently. We therefore come to the conclusion that the fiber impulse response, calculated or measured, is only meaningful under well defined conditions.

3.6 Graded Index Fibers

A quick look at the impulse responses of the previous section will make it abundantly clear that multimode step-index fibers are rather dispersive and not particularly suitable for broadband communications. We saw in Sect.3.5 that a suitably graded waveguide behaves as a lens-like medium, which equalizes the various optical paths. Thus, we would expect that some improvement in the dispersion properties should be possible if the fiber were to be graded. This is indeed the feature responsible for the present day importance of graded index fibers. In this section, we use a similar method as in Sect.3.4, and derive the eigenvalue equation for such fibers, and then proceed to calculate the impulse response. In our analysis, we will assume that the refractive index of the fiber can be described by the so called α profile [3.39], given by

$$n^2(\rho) = n_0^2[1 - 2\Delta(\tfrac{\rho}{a})^\alpha] \quad , \quad 0 < \rho < a \quad , \tag{3.111}$$

where n_0 is the refractive index at $\rho = 0$, and $\Delta \approx [n_0 - n(a)]/n_0$. This representation is rather convenient because profiles ranging from triangular to rectangular can be obtained simply by varying α between 1 and ∞. For example, $\alpha = 2$ gives us the parabolic profile.

The analysis of graded index fibers is based on the vector wave equation (3.58), from which the z component of the vector Laplacian separates to give the scalar Helmholtz equation (Sect.3.5). Separation of variables leads to the same form of radial equations as in Sect.3.5, with the constant n replaced by $n(\rho)$. In other words, the radial equation becomes

$$\frac{d^2F}{d\rho^2} + \frac{1}{\rho}\frac{dF}{d\rho} + [n^2(\rho)k_0^2 - \beta^2 - \nu^2/\rho^2]F = 0 \quad . \tag{3.112}$$

We will first solve this equation around the turning points ρ_1 and ρ_2 using the method of Sect.3.4. To be able to use the Airy function, (3.112) must first be transformed by setting $F(\rho) = f(\rho)/\sqrt{\rho}$. By direct substitution, the differential equation for $f(\rho)$ becomes

$$\frac{d^2f}{d\rho^2} + [n^2(\rho)k_0^2 - \beta^2 - \frac{(\nu^2-1/4)}{\rho^2}]f(\rho) = 0 \quad . \tag{3.113}$$

For any ρ_1 and ρ_2 ($\neq 0$) the radial propagation constant k_ρ^2 can be expressed in terms of the Taylor series as in (3.75)

$$n^2(\rho)k_0^2 - \beta^2 - \frac{(\nu^2-1/4)}{\rho^2} = -k_0^2N_1(\rho - \rho_1) + O[(\rho - \rho_1)^2] \quad , \tag{3.114}$$

where now

$$N_1 = -2n(\rho_1)\frac{dn(\rho_1)}{d\rho} + \frac{2(\nu^2-1/4)}{\rho_1^3k_0^2} \quad . \tag{3.115}$$

As in Sect.3.4, we obtain solutions of the form

$$f(\rho) = f_1\mathrm{Ai}\left[k_0^{2/3}\,N_1^{1/3}(\rho - \rho_1)\right] \quad . \tag{3.116}$$

with asymptotic solutions of the form given by (3.77) and (3.78). For ρ_2 analogous results are obtained.

For $f(\rho)$, the WKB-method works as in Sect.3.4, but now the phase function must be defined as

$$G(\rho) = \sqrt{n^2(\rho) - \frac{\beta^2}{k_0^2} - \frac{(\nu^2-1/4)}{k_0^2\rho^2}} \quad . \tag{3.117}$$

The radial solutions around ρ_1 then become

$$F(\rho) = \frac{F_1N_1^{1/6}\,k_0^{1/3}}{2\sqrt{\pi\rho\,|G(\rho)|}}\exp\left[-k_0\int_{\rho_1}^{\rho}|G(\rho')|d\rho'\right] \quad , \quad \rho > \rho_1 \quad , \tag{3.118}$$

and,

$$F(\rho) = \frac{F_1 N_1^{1/6} k_0^{1/3}}{\sqrt{\pi\rho G(\rho)}} \sin\left[k_0 \int_\rho^{\rho_1} G(\rho')d\rho' + \frac{\pi}{4}\right] \quad , \quad \rho < \rho_1 \quad . \tag{3.119}$$

Similarly, around ρ_2 the solutions become:

$$F(\rho) = \frac{F_2 N_2^{1/6} k_0^{1/3}}{2\sqrt{\pi\rho|G(\rho)|}} \exp\left[-k_0 \int_\rho^{\rho_2} |G(\rho')|d\rho'\right] \quad , \quad 0 \leqq \rho < \rho_2 \quad , \tag{3.120}$$

$$F(\rho) = \frac{F_2 N_2^{1/6} k_0^{1/3}}{\sqrt{\pi\rho G(\rho)}} \sin\left[k_0 \int_{\rho_2}^{\rho} G(\rho')d\rho' + \frac{\pi}{4}\right] \quad , \quad \rho > \rho_2 \quad . \tag{3.121}$$

Because of the uniqueness of the field for $\rho_2 < \rho < \rho_1$ the solutions given by (3.119) and (3.121) must be identical so that

$$k_0 \int_{\rho_2}^{\rho_1} G(\rho)d\rho = (\mu - \tfrac{1}{2})\pi \quad . \tag{3.122}$$

It should be noted that the solutions given by (3.120) and (3.121) are approximate, also in the sense that for the inner cut-off region the other Airy function Bi(.) also exists. For further use we write (3.122) in the form

$$\int_{\rho_2}^{\rho_1} \sqrt{k_0^2 n^2(\rho) - \beta^2 - (\nu^2 - 1/4)/\rho^2}\, d\rho = (\mu - \tfrac{1}{2})\pi \quad . \tag{3.123}$$

The solution of this equation gives the propagation constant $\beta(\mu,\nu)$ for any modal integers μ (radial variation) and ν (azimuthal variation). This forms the eigenvalue equation of the graded index fiber.

We see that the function under the integral sign remains real for $[k_0^2 n^2(\rho) - \beta^2] > (\nu^2 - 1/4)/\rho^2$, and is imaginary otherwise. The former corresponds to propagating solutions, while the latter to decaying solutions. Figure 3.37 shows this situation graphically for two different values of ν. Figure 3.37a corresponds to a propagating mode which has oscillating solutions between ρ_1 and ρ_2, and decaying solutions beyond. On the other hand, Fig.3.37b shows a mode which has propagating solutions between ρ_1 and ρ_2, a non-propagating region between ρ_2 and ρ_3, and once again propagating solutions beyond ρ_3. We see that, as in Sect.3.2 and 3.5, we have a situation in which the wave "tunnels" through a forbidden region and re-appears beyond a certain

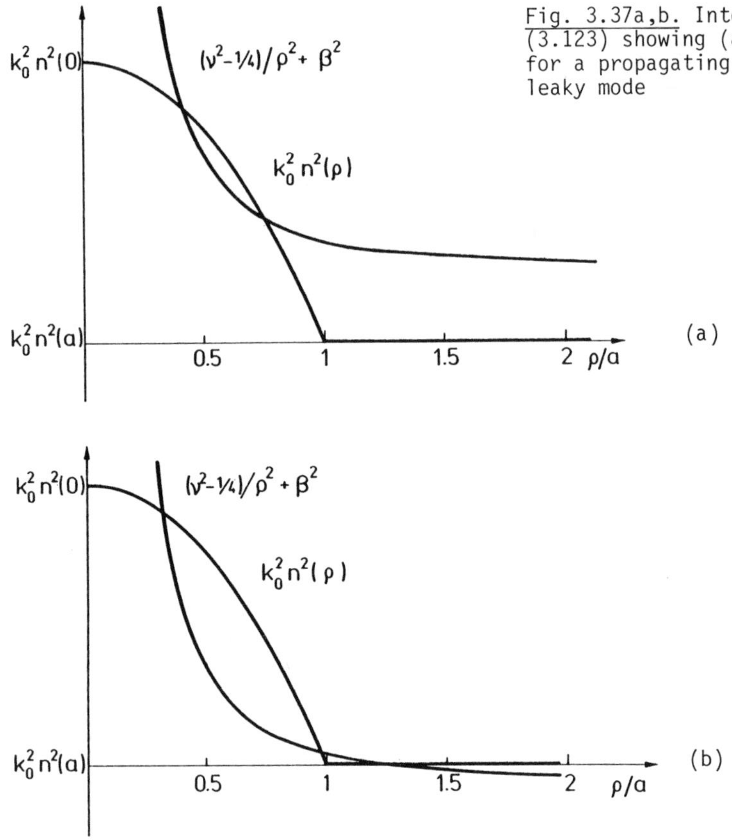

Fig. 3.37a,b. Interpretation of (3.123) showing (a) the condition for a propagating mode, and (b) a leaky mode

point (ρ_3). In other words, all μ,ν combinations which lead to this situation, represent leaky modes of the graded index fiber. Such modes gradually pump their energy from the core region into the cladding, and cannot be true propagating modes.

Let us now examine the behaviour of the eigenvalue equation (3.123) for two specific index profiles.

Example 1. Equation (3.123) is easiest to evaluate for the step index case since the core and cladding refractive indices are then constant, i.e. $n(\rho) = n_1$ for $\rho < \rho_1$, and $n(\rho) = n_2$ for $\rho < \rho_1$. Thus, for guided modes, we have:

$$\int_{\rho_2}^{a} d\rho \sqrt{k_0^2 n_1^2 - \beta^2 - \frac{(\nu^2-1/4)}{\rho^2}} = (\mu - \frac{1}{2})\pi \qquad (3.124)$$

$$= \nu_1\{\tan[\arccos(\nu_1/u)] - \arccos(\nu_1/u)\} \quad , \tag{3.124}$$

where

$$\rho_2 = \sqrt{\nu^2 - 1/4} \Big/ \sqrt{n_1^2 k_0^2 - \beta^2} \quad , \quad u = a\sqrt{k_0^2 n_1^2 - \beta^2} \quad ,$$

and $\nu_1 = \sqrt{\nu^2 - 1/4}$. With some manipulation, the following result is obtained:

$$\mu = \frac{1}{2} + \frac{1}{\pi}\left(\sqrt{u^2 - \nu_1^2} - \nu_1 \arccos \frac{\nu_1}{u}\right) \quad . \tag{3.125}$$

This result can be presented in the form of a mode diagram as shown in Fig. 3.38. Cut-off for u is attained at $u = V = ak_0\sqrt{n_1^2 - n_2^2}$. Note that the maximum value of ν_1 is also V.

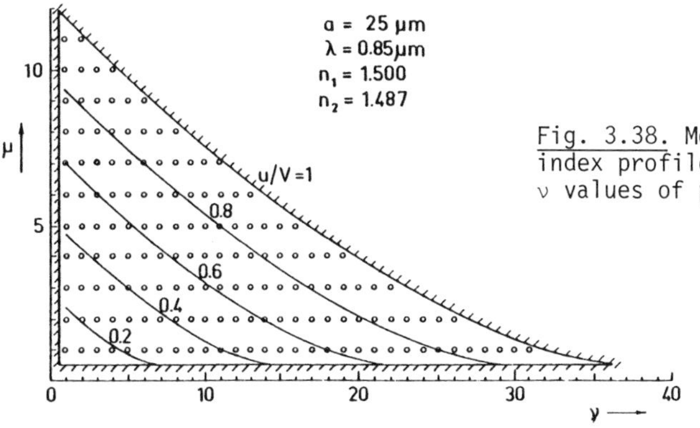

Fig. 3.38. Mode diagram for a step-index profile showing the allowed μ, ν values of propagating modes

Example 2. For the parabolic index profile, α in (3.111) equals 2, and (3.123) becomes:

$$\int_{\rho_2}^{\rho_1} \left[k_0^2 n^2(0) - \beta^2 - 2k_0^2 \frac{n^2(0)\Delta}{a^2}\rho^2 - \frac{\nu_1^2}{\rho^2}\right]^{\frac{1}{2}} d\rho = (\mu - \tfrac{1}{2})\pi \tag{3.126}$$

$$= \int_{\rho_2}^{\rho_1} \frac{d\rho}{\rho}\sqrt{\frac{u^2\rho^2}{a^2} - \frac{V^2\rho^4}{a^2} - \nu_1^2} \quad , \quad \text{where}$$

now $u = a\sqrt{k_2^2 n^2(0) - \beta^2}$, $V = ak_0 n(0)\sqrt{2\Delta}$, and $\nu_1 = \sqrt{\nu^2 - 1/4}$. The limits ρ_1

and ρ_2 are the radii for which the term in the brackets vanishes. If we now change the variable ρ to $W = \rho^2 V/a\nu_1$, the integral in (3.126) becomes

$$\frac{\nu_1}{2}\int_{W_2}^{W_1}\frac{dW}{W}\sqrt{\frac{u^2}{V\nu_1}W - W^2 - 1} = \frac{\nu_1}{2}\int_{W_2}^{W_1}\frac{dW}{W}\sqrt{-W^2 + (W_1 + W_2)W - W_1W_2} \quad , \tag{3.127}$$

where

$$W_{1,2} = \frac{u^2}{2V\nu_1}\left(1 \pm \sqrt{1 - \frac{4V^2\nu_1^2}{u^4}}\right) \quad . \tag{3.128}$$

This integral is tabulated (see e.g. [3.40]) and reduces to

$$\frac{\nu_1}{2}\left[\sqrt{(W - W_2)(W_1 - W)} - \arcsin\left(\frac{-2W_1W_2/W + W_1 + W_2}{W_1 - W_2}\right)\right.$$

$$\left.- \frac{(W_1 + W_2)}{2}\arcsin\left(\frac{-2W + W_1 + W_2}{W_1 - W_2}\right)\right]_{W_2}^{W_1} = \frac{\nu_1}{2}\left(0 - \pi + \frac{u^2\pi}{2V\nu_1}\right) = (\mu - \frac{1}{2})\pi \quad . \tag{3.129}$$

We re-write this in terms of μ in order to give it the same form as (3.125), i.e.

$$\mu = \frac{1}{2} + \frac{u^2}{4V} - \frac{\nu_1}{2} \quad . \tag{3.130}$$

The contours of this mode characteristic are shown in Fig.3.39, and it can be seen that they are now almost linear. The cut-off for u is still at V, while the maximum value for $\nu_1 = V/2$, and for $\mu = V/4$.

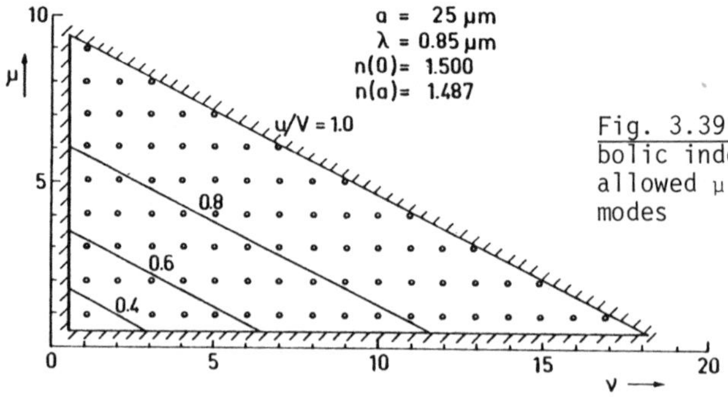

Fig. 3.39. Mode diagram for a parabolic index profile showing the allowed μ, ν values of propagating modes

Let us now return to the general α profile and evaluate the number of modes which can be supported by a fiber. To do this, we let $m(\beta)$ denote the number of modes with $\beta \geq k_0 n(\rho)$, for any ν. Then we set $\nu \approx \nu_1$ and use it as a continuous variable so that from (3.123) we obtain:

$$m(\beta) = \frac{4}{\pi} \int_0^{\nu_{max}} d\nu \int_{\rho_2(\nu)}^{\rho_1(\nu)} d\rho \sqrt{k_0^2 n^2(\rho) - \beta^2 - \frac{\nu^2}{\rho^2}}$$

$$= \frac{4}{\pi} \int_0^{\rho_1} d\rho \int_0^{\rho\sqrt{k_0^2 n^2 - \beta^2}} d\nu \sqrt{k_0^2 n^2(\rho) - \beta^2 - \nu^2/\rho^2} \quad , \tag{3.131}$$

where the factor 4 accounts for the fact that each (μ,ν) mode may have two polarizations and two directions of rotation. The lower expression has been derived from the upper one by changing the order of integration. The integral over ν can be easily evaluated if we replace β by $k_0 n(\rho_1)$; then (3.131) reduces to the following

$$m(\beta) = \int_0^{\rho_1} \rho d\rho \sqrt{k_0^2 n^2(\rho) - \beta^2} \quad . \tag{3.132}$$

This is rather convenient, because the above integral can be evaluated analytically for the α profile. Thus, if we take

$$\rho_1 = a\left[\frac{1}{2\Delta}\left(1 - \frac{\beta^2}{k_0^2 n^2(0)}\right)\right]^{1/\alpha} = a\left(\frac{u}{V}\right)^{2/\alpha} \quad , \tag{3.133}$$

then

$$m(\beta) = \frac{\alpha}{\alpha + 2} a^2 k_0^2 n^2(0)\Delta \left[\frac{1}{2\Delta}\left(1 - \frac{\beta^2}{k_0^2 n^2(0)}\right)\right]^{1+2/\alpha}$$

$$= \frac{\alpha}{(\alpha + 2)} \frac{V^2}{2} \left(\frac{u}{V}\right)^{2+4/\alpha} \quad . \tag{3.134}$$

The above expression is displayed in Fig.3.40 as a function of u, for several values of α, and we see that maximum $m(\beta_c) = M$ is obtained for $\beta_c = k_0 n(a)$, or $u = V$. Thus, the total number of modes (M) is given by

$$M = m(\beta_c) = \frac{\alpha}{\alpha + 2} a^2 k_0^2 n^2(0)\Delta = \frac{\alpha}{\alpha + 2} \frac{V^2}{2} \quad . \tag{3.135}$$

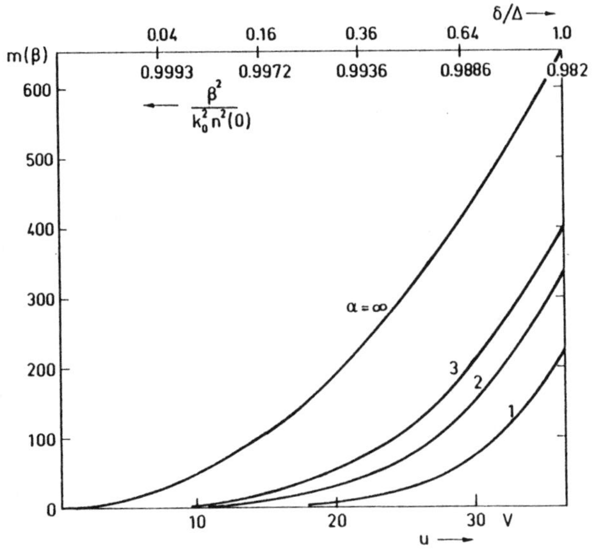

Fig. 3.40. Graphical illustration of (3.134) for various values of α

We see that $M = V^2/4$ for the parabolic profile ($\alpha = 2$), while $M = V^2/2$ for the step-index fiber ($\alpha = \infty$). Furthermore, (3.134) and (3.135) show that the number of modes increases quadratically with respect to V and even faster with respect to u. Thus, individual mode control is rather difficult.

Let us now turn to the calculation of modal delay. Firstly, we note that radial field solutions (3.120) and (3.121), together with the azimuthal solutions $[\exp(\pm j\nu\phi)]$, and the axial solutions $[\exp(\pm j\beta z)]$, give the complete picture of the longitudinal waveguide field components E_z and H_z. The perpendicular components are then obtained by partial derivation. Thus the modal picture becomes complete when the initial boundary conditions of mode excitation are given. However, this is not usually attempted for multimode waveguides. Rather, it is assumed that the individual modes add incoherently in power, and that the delay of each mode can be computed from (3.86). We therefore re-write this equation by taking k_0 and β as the independent variables [tied by (3.123)], and $\mu = \mu(k_0,\beta)$ as the dependent variable. Equation (3.86) then becomes

$$\tau(\mu,\nu) = -\frac{1}{c}\frac{\left(\frac{\partial\mu}{\partial k_0}\right)_\beta}{\left(\frac{\partial\mu}{\partial\beta}\right)_{k_0}} \quad . \tag{3.136}$$

This equation will be our starting point for the calculation of the propagation delay of each mode (and hence the impulse response of the fiber). Its partial derivatives can be obtained from the eignvalue equation (3.123). Thus, representing the radial component of the wave vector by

$$U(\rho) = k_0 G(\rho) = \sqrt{k_0^2 n^2(\rho) - \beta^2 - \frac{(\nu^2 - 1/4)}{\rho^2}} \quad , \tag{3.137}$$

we obtain

$$\frac{\partial\mu}{\partial k_0} = \frac{1}{\pi}\int_{\rho_2}^{\rho_1} \frac{\partial U}{\partial k_0} d\rho = \frac{1}{\pi}\int_{\rho_2}^{\rho_1} \frac{k(\rho)N(\rho)}{U(\rho)} d\rho \quad , \tag{3.138}$$

$$\frac{\partial\mu}{\partial\beta} = -\frac{\beta}{\pi}\int_{\rho_2}^{\rho_1} \frac{d\rho}{U(\rho)} \quad . \tag{3.139}$$

Here $k(\rho) = k_0 n(\rho)$. In order to account for the wavelength dependence of the refractive index, we have used the group index $N(\rho)$ given by

$$N(\rho) = n(\rho) + k_0 \frac{\partial n(\rho)}{\partial k_0} = n(\rho) - \lambda \frac{\partial n(\rho)}{\partial\lambda} = \frac{\partial k}{\partial k_0} \quad . \tag{3.140}$$

We digress a little and note that the above partial derivatives also allow a differential geometric interpretation. Thus, referring to Fig.3.41, consider the ray which propagates in the direction $\underline{k}(\rho)$. The directional cosines of the ray can be expressed by the curved length element ds, so that

$$\frac{d\rho}{ds} = \frac{U(\rho)}{k(\rho)} \quad , \quad \text{and} \quad \frac{dz}{ds} = \frac{\beta}{k(\rho)} \quad . \tag{3.141}$$

Then using (3.138,139 and 140) we obtain

$$\tau = \frac{\int_{s_2}^{s_1} ds N(\rho)/c}{\int_{z_2}^{z_1} dz} \quad , \tag{3.142}$$

where the coordinates (s_2, s_1), (z_2, z_1) refer to the turning points of the ray. The ray traverses from the inner caustic at $\rho = \rho_2$ to the outer caustic

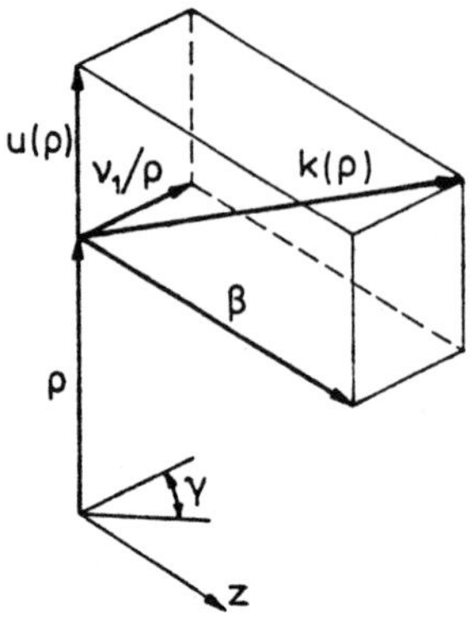

Fig. 3.41. Differential-geometric interpretation afforded by the WKB-method

at $\rho = \rho_1$ and turns around helically at the same time. The numerator measures the time spent, and the denominator indicates the axial distance covered.

Let us now return to the main theme. In order to apply (3.136) (for the α profile) we re-write (3.123) in a form similar to that in (3.126)

$$\mu - \frac{1}{2} = \frac{1}{\pi}\int_{\rho_1}^{\rho_2} d\rho\left[\frac{u^2}{a^2} - \frac{V^2}{a^2}\left(\frac{\rho}{a}\right)^{\alpha} - \frac{\nu_1^2}{\rho^2}\right]^{\frac{1}{2}}$$

$$= \frac{\nu_1}{2\pi}\int_{W_1}^{W_2} \frac{dW}{W}\left(u^2 W \nu_1^{-2\alpha/(2+\alpha)}\, V^{-4/(2+\alpha)} - W^{1+\alpha/2} - 1\right)^{\frac{1}{2}} , \qquad (3.143)$$

where $W = (\rho/a)^2 (V/\nu_1)^{4/(2+\alpha)}$. Now denote the integrand for a moment with $f(u,V,W)$, where $V = \sqrt{2\Delta}\, k_0 n(0)a$. The bounds of the integral are the two values at which $f(u,V,W) = 0$. The differentiation can be performed, and only the part shown below is non-zero

$$\tau = \frac{1}{c}\left(\frac{\partial\beta}{\partial k_0}\right)_{\mu} = -\frac{1}{c}\,\frac{\int_{W_1}^{W_2} dW\, \frac{\partial f(u,V,W)}{\partial k_0}}{\int_{W_1}^{W_2} dW\, \frac{\partial f(u,V,W)}{\partial \beta}} . \qquad (3.144)$$

The differentiation of $f(u,V,W)$ is quite tedious and results in

$$\frac{\partial f}{\partial k_0} = \frac{\partial f}{\partial u}\frac{\partial u}{\partial k_0} + \frac{\partial f}{\partial V}\frac{\partial V}{\partial k_0} =$$

$$= \frac{\nu_1^{-2\alpha/(2+\alpha)} V^{-4/(2+\alpha)}}{Wf(u,V,W)} \left[a^2 n(0) N(0) k_0 - \frac{(2+\zeta/2)u^2}{(2+\alpha)k_0} \frac{N(0)}{n(0)} \right] , \tag{3.145}$$

$$\frac{\partial f}{\partial \beta} = \frac{\partial f}{\partial u} \frac{\partial u}{\partial \beta} = \frac{\nu_1^{-2\alpha/(2+\alpha)} V^{-4/(2+\alpha)}}{Wf(u,V,W)} \left(-\frac{a^2 \beta}{u} \right) , \tag{3.146}$$

where N(0) is the group index at the axis, and

$$\zeta = - \frac{2n(0)}{N(0)} \frac{\lambda}{\Delta} \frac{d\Delta}{d\lambda} = - \frac{2n(0)}{N(0)} \frac{d(\ln\Delta)}{d(\ln\lambda)} . \tag{3.147}$$

It is seen that the integrals over $[Wf(u,V,W)]^{-1}$ cancel and we obtain

$$\tau = \frac{N(0)}{c} \left[\frac{k_0}{\beta} - \frac{(2+\zeta/2)u^2}{(2+\alpha)k_0 n(0) \beta a^2} \right]$$

$$= \frac{N(0)}{c} \frac{1 - \delta(4+\zeta)/(\alpha+2)}{\sqrt{1-2\delta}} , \tag{3.148}$$

where the variable $\delta = \delta\,(\beta, k_0) \in [0,\Delta]$ is given by

$$\delta = \frac{1}{2} \left[1 - \frac{\beta^2}{k_0^2 n^2(0)} \right] = \left(\frac{u}{V} \right)^2 \Delta . \tag{3.149}$$

The expression in (3.148) can be modified to give the normalized differential delay (for any δ) as follows

$$t_\delta = \frac{\tau - N(0)/c}{N(0)/c} = \frac{\tau c}{N(0)} - 1$$

$$= \left[1 - \frac{(4+\zeta)\delta}{\alpha+2} \right] \left[1 + \delta + \frac{3\delta^2}{2} + \dots \right] - 1$$

$$= \frac{(\alpha - 2 - \zeta)}{(\alpha+2)} + \frac{(3\alpha - 2 - 2\zeta)}{(\alpha+2)} \frac{\delta^2}{2} + \frac{(5\alpha - 2 - 3\zeta)}{(\alpha+2)} \frac{\delta^3}{2} + \dots . \tag{3.150}$$

In particular, for $\alpha = 2 + \zeta$ the delay t_δ becomes a quadratic function of δ [scaled radial component squared according to (3.143)]:

$$t_\delta = \frac{\delta^2}{2} (1 + 2\delta + \dots) . \tag{3.151}$$

For $\alpha \neq 2+\zeta$, the series in (3.150) can be inverted to give

$$\delta = \frac{t_\delta(\alpha+2)}{\alpha-2-\zeta}\left[1 - \frac{t_\delta}{2}\frac{(\alpha+2)(3\alpha-2-2\zeta)}{(\alpha-2-\zeta)^2} + \dots\right] \quad . \tag{3.152}$$

For $\alpha = 2+\zeta$ a singular solution is reached

$$\delta \doteq \sqrt{2t_\delta}(1 - \sqrt{2t_\delta} + \dots) \quad . \tag{3.153}$$

Armed with this information, we can now derive an explicit expression for the fiber impulse response. Consider that for a certain delay t_δ, corresponding to the propagation constant β, the amount of energy received is proportional to the number of modes (μ,ν) in the group. Thus, in an interval Δt_δ, this number must be given by the corresponding $\Delta m(\beta)$. The impulse response is then approximately $\Delta m(\beta)/\Delta t_\delta$ and can be normalized to unity by dividing by M, the total number of modes. (We have assumed in the above arguments that the modes add incoherently in power). For a sufficiently large number of modes, we can write the impulse response in terms of derivatives. Then, use of (3.134) and (3.152) yields the impulse response as follows

$$\begin{aligned} h(t) &= \frac{1}{M}\frac{dm}{dt_\delta} = \frac{1}{M}\frac{d}{dt_\delta}\left[M\left(\frac{\delta}{\Delta}\right)^{1+2/\alpha}\right] \\ &\doteq \frac{1}{\Delta^{1+2/\alpha}}\frac{d}{dt_\delta}\left[\left(\frac{\alpha+2}{\alpha-2-\zeta}\right)^{1+2/\alpha} t_\delta^{1+2/\alpha}\right] \\ &= \left|\frac{\alpha+2}{\alpha-2-\zeta}\,\frac{1}{\Delta}\right|^{1+2/\alpha}\left(1+\frac{2}{\alpha}\right)t_\delta^{2/\alpha} \quad . \end{aligned} \tag{3.154}$$

Note that only the first term in (3.152) (for $\alpha \neq 2+\zeta$) has been used here. Similarly, for $\alpha = 2+\zeta$, the impulse response is approximately

$$h(t) \doteq \frac{2}{\Delta^2} \quad . \tag{3.155}$$

The impulse response and its corresponding α-profile are shown in Fig.3.42. To convert the results into real time, the time scale should of course be multiplied by N(0)/c.

The width of the impulse response can be obtained from (3.150) or (3.151) by setting $\delta = \Delta$. This yields:

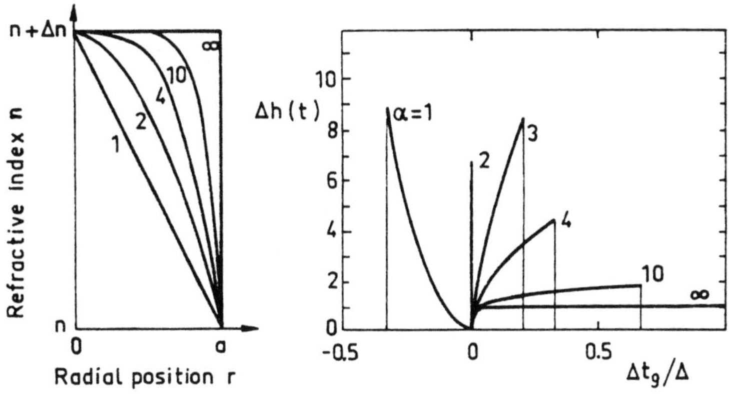

(a) (b)

Fig. 3.42. Refractive index profile shapes (a) for various values of α, and the corresponding impulse responses (b)

$$T = t_\Delta \doteq \left|\frac{\alpha - 2 - \zeta}{\alpha + 2}\right| \Delta \quad , \qquad \text{for } \alpha \neq 2 + \zeta \quad , \tag{3.156}$$

and,

$$T \doteq \frac{\Delta^2}{2} \quad , \qquad \text{for } \alpha = 2 + \zeta \quad . \tag{3.157}$$

Thus for $\Delta = 8.9 \cdot 10^{-3}$, $N(0) = 1.5$, and $\zeta = 0$, the impulse response will have the widths given in Table 3.1. Clearly, a well compensated index profile can lead to a very small dispersion. The optimum profile corresponds to $\alpha = 2 - 2\Delta + \zeta$, and is simply obtained as the approximate zero of the first two terms in (3.150).

Table 3.1. Impulse response width as a function of the fiber parameter α

α	T [ns/km]
1	14.7
2	0.201
3	9.2
∞	44.9

We conclude our analysis by summarizing the following essential aspects of this section:

1) The WKB approximation breaks down at the turning points or caustics, and a better approximation is provided (at the caustics) by Airy functions. We saw that an acceptable field description can be obtained by matching WKB and asymptotic Airy solutions. This technique provided us with a convenient derivation of the eigenvalue or characteristic equation of the graded index fiber.

2) We saw that the impulse response can be analytically calculated using the α-profile approximation. The analytical expressions for the impulse response and its width, made it clear that an optimum profile exists. Although this is important information, and the analysis on the whole provides us with good physical insight into fiber behaviour, a word of caution is necessary! It has been shown e.g. by RAMSKOV-HANSEN [3.41] that the α-profile approximation is rather poor in a number of ways. Furthermore, we saw in Sect.3.5, that certainly for $\alpha = \infty$ (step index case), the WKB approximation itself deviates considerably from the exact solution.

3) As already stated at the end of Sect.3.5, we should remember that launching conditions and mode mixing may, in practice, give rise to impulse responses that differ significantly from the calculated ones.

4) Finally, we remind the reader that use of the Helmholtz equation for the graded index case is only valid provided the conditions of (3.59) and (3.60) can be fulfilled.

3.7 Coherence Effects and Speckle in Fibers

One of the easiest ways to show the existence of different modes in fibers is to coherently illuminate the input end of the fiber (Fig.3.43), and to observe the existence of bright and dark spots in the far-field pattern of the fiber. In the terminology of coherent optics, such a picture is known as a "speckle pattern" [3.42].

In this section, we would like to briefly review the nature and main properties of speckle, and then use these considerations to explain speckle phenomena in fibers. Consider therefore, the situation shown in Fig.3.44a which shows a surface illuminated either by source L_1 (for "reflecting" surface) or source L_2 (for semitransparent surface). Each small area of the rough

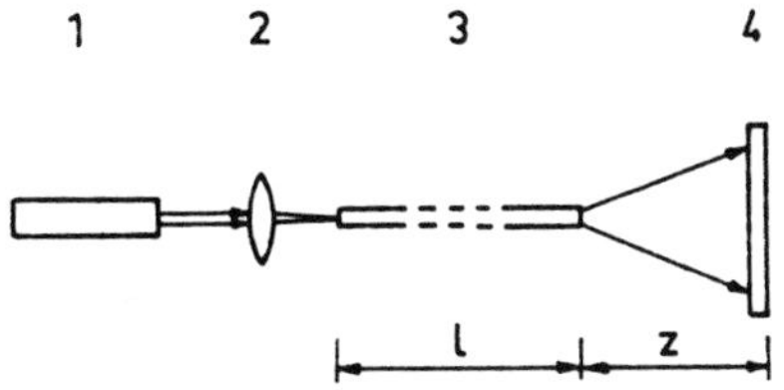

Fig. 3.43. Experimental set-up for the observation of fiber speckle: (1) laser, (2) launching lens, (3) fiber, (4) scattering plate

Fig. 3.44. Set-ups for the observation of speckle phenomena (a) in the far field, and (b) in an image produced by a lens

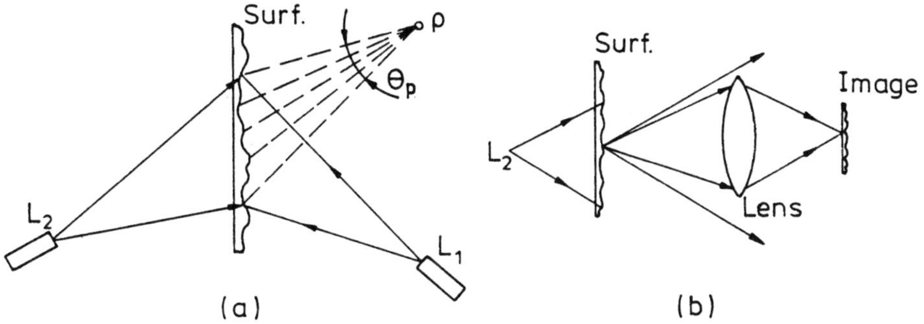

surface becomes an induced source of scattered radiation, and depending on the degree of roughness of the surface (in comparison to the wavelength), the phase differences amongst these sources may vary from a few to thousands of radians. Thus, at every observation point such as P, the field intensity will be determined by the mutual interference of a large number of coherent but strongly de-phased wavelets. If the observation point is moved, the path lengths of the scattered components will change, and a new and independent intensity distribution will result. Consequently, the far-field pattern will be a random sequence of bright (constructive interference) and dark (destructive interference) "spots".

We must stress here that speckle is not caused by the imaging system (the eye, for example), as can be easily shown by exposing a photographic plate to the far-field radiation, and by consequent microscopic observation of the pattern obtained. When image formation elements are used (Fig.3.44b) one might expect that speckle will not occur because rays from a given small area will be focused to their corresponding image. However, we also obtain speckle in this case because all rays cannot be focused in precise correspondence. Even for aberration-free optics, a limited aperture may lead to incomplete phase information at the image, and produce abrupt changes in the resulting intensity. In fact, the larger the aperture of the focusing lens, the more complete is the phase information, and the smaller the influence of speckle. However, it is only possible to suppress the influence of speckle on image quality when the angular aperture of the lens is large enough to gather all scattered radiation from the illuminated surface.

GOODMAN [3.42,43] alloted two important features to the elementary scatterers responsible for speckle formation: (I) that the amplitudes and phases of the elementary wavelets are statistically independent, and (II) that the phases of elementary contributions are equally likely to lie anywhere in the interval $(-\pi, +\pi)$. (The latter is equivalent to assuming a surface which is rough in comparison to a wavelength). Based on these assumptions, it is possible to show that the irradiance I of the speckle pattern must follow negative exponential statistics, i.e. its probability density function must be of the form:

$$\begin{aligned} P(I/\bar{I}) &= \exp(-I/\bar{I}) \quad , \qquad I/\bar{I} \geqq 0 \\ &= 0 \qquad\qquad\qquad \text{otherwise} \quad , \end{aligned} \tag{3.158}$$

where $\bar{I}$ is the mean or expected irradiance. This function is shown by the solid curve of Fig.3.45 for the polarized speckle pattern. A fundamental feature of this distribution is that its mean and standard deviation are identical, so that the contrast ratio (standard deviation: mean) of a polarized speckle pattern is always unity. Herein lies the reason for the subjective impression of the high contrast of speckle patterns.

Let us next consider the coarseness of grains in speckle patterns. Firstly, we note that any two-dimensional intensity distribution can be considered in terms of a Fourier series of sinusoidal components of different amplitudes and periods. The amplitude spectrum which is usually presented as a Wiener spectrum can then be used to obtain a measure of the average grain size. GOODMAN defined the average size G to be the minimum period present in the Wiener spectrum. Referring to Fig.3.46, we see that the grain size is given by

$$G = \frac{\lambda z}{L} \quad , \tag{3.159}$$

where λ is the wavelength, L is the length of the side of a square scattering area, and z is the distance to the observation point.

Each sinusoidal intensity component can be considered to be due to the interference between two plane waves. As is well known, the period of the intensity pattern is then approximately given by

$$P \simeq \frac{\lambda}{\theta} \tag{3.160}$$

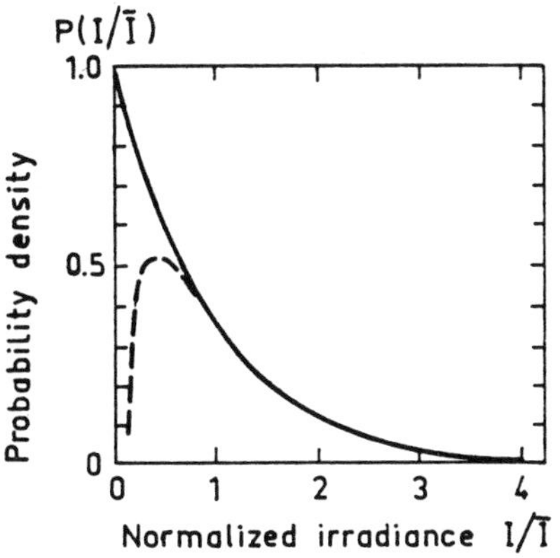

Fig. 3.45. Probability densities for the polarized speckle of an ideal scatterer (——) and a coherently illuminated fiber (----) [3.46]

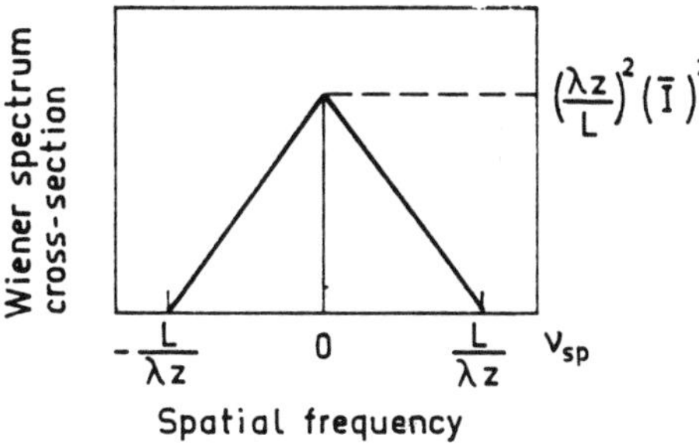

Fig. 3.46. A sketch of the Wiener spectrum of an ideal speckle pattern [3.43]

for small θ, where θ is the angle subtended at the observation point by the two plane waves. Moreover, reference to Fig.3.44a shows that, if θ is the angle defined by the two marginal rays (marked solid), the sinusoid generated by these two rays corresponds to the minimum period in the Fourier series

$$P_{min} \simeq \frac{\lambda}{\theta} \simeq \frac{\lambda z}{L} \quad . \tag{3.161}$$

Thus, we see that the grain size given in (3.159) is the same as this minimum period, since θ is maximum. The important feature to notice is that grain size decreases with increasing θ. In other words, the further away is the observation point, the larger will be the grain size.

Proceeding to the fiber case, we recall that the main reason for speckle in fiber optics is the existence of various modes, each with a different propagation constant and therefore different group velocity (Sect.3.5). As a result, if the fiber input is excited coherently as shown in Fig.3.43, each mode will have a specific amplitude and phase distribution at the output end of the fiber, and we would expect different modes to interfere with each other. However, for such interference to occur, the propagation delay difference amongst modes must remain less than the temporal coherence of the source. The group velocity of a given mode in a weakly guiding step index fiber (Sect.3.5) is given by [3.44,45]:

$$v_g(\omega,\ell) \simeq \left(1 - \ell^2\Delta/\ell_m^2\right)c/n_1 \quad , \tag{3.162}$$

where ℓ is an integer varying from 2 to ℓ_m, and ℓ_m is given by

$$\ell_m \simeq 2an_1(2\Delta)^{\frac{1}{2}}/\pi \quad . \tag{3.163}$$

Here, a is the core radius, n_1 is the group refractive index of the core, and Δ is the fractional core-cladding index difference. Thus, the delay difference between the slowest and fastest mode, for a fiber of length z, is given by

$$T_1 \simeq n_1 z\Delta/c \quad . \tag{3.164}$$

On the other hand, the minimum delay difference (T_2) occurs between modes with $\ell = 2$ and $\ell = 3$, and can be easily found to be

$$T_2 = 5\Delta zn_1/(\ell_m^2 c) \quad . \tag{3.165}$$

These two delay differences should then be compared to the coherence time of the source, given by

$$T_s = 2\pi/\delta\omega \quad , \tag{3.166}$$

where $\delta\omega$ is the spectral width of the source. Thus, for $T_2 > T_s$, no interference can occur, while for $T_1 > T_s$ some modes may still be able to interfere. Furthermore, observe that increasing the spectral width of the source decreases the coherence time.

In any case, we see that, when intermodal temporal coherence exists, the phase will vary across the fiber end face, in much the same way as across the rough surface of Fig.3.44. However, in the fiber case, the intensity will also be different at different points. Moreover, experiments show that, for multimode fibers, a length of 10 to 20 cm is sufficient for the depolarization of the input wave. Thus, the fiber end face is also different in this respect from the rough surface. However, from the viewpoint of speckle, we can still consider the fiber end face as an equivalently illuminated surface of the same size. Under these assumptions, we would expect the fiber to produce speckle with the same statistics as discussed in the earlier parts of this section. The dashed curve in Fig.3.45 shows the probability density function of speckles in the fiber far-field [3.46]. In this case, we see that negative exponential statistics are only applicable over a limited region, so that we would expect a decrease in contrast. A possible reason for this deviation may be that classical speckle laws assume a homogeneously illuminated surface, while the fiber end face may be quite unevenly illuminated.

We conclude this section by observing that speckle properties depend not only on fiber length, core diameter, and mode structure but also on such factors as the number of leaky and cladding modes, as well as on mode coupling effects. As a result, future investigations of coherence effects in fiber may lead to interesting methods for evaluating these properties, and for their use in sensor devices. However, as will be seen in Sect.6.2, the existence of speckles places a major restriction on the viability of certain types of analog systems.

4. Components for Optical Fiber Systems

In Chap.2 we reviewed some of the physics of light generation, modulation, and detection, while in Chap.3 we considered the behavior of light as it propagates through the fiber channel. We should now familiarize ourselves with the structure of the actual components that form a communication system, as well as consider the means for coupling them together. Thus, we first review some methods for the fabrication of fibers, and then discuss sources, modulators, and detectors. Coupling problems are treated in Sect.4.5.

4.1 Fiber Fabrication Methods

While the concept of a dielectric waveguide is not new and was already discussed in 1910 by HONDROS and DEBYE [3.34], from the viewpoint of communications, hope and interest were in fact triggered in 1966 when KAO and HOCKHAM [4.1] published the results of their research. They realized that a major reason for the customarily high attenuations (1-10 dB/m) observed in those days was the existence of undesirable impurities in the glasses, and that the attenuation could be reduced to acceptable levels by the use of sufficiently pure materials. They were proved correct in 1970, when, using the ultra-pure ingredients commonly found in the semiconductor industry, the group at Corning Glass produced a low-loss fiber (20 dB/km) using the "flame-hydrolysis" method. The explosive development that followed is evident in the attenuation figures of 5 dB/km in 1972, 1 dB/km in 1977, and 0.2 dB/km in 1979 (albeit at 1.55 μm). At the same time, bandwidths of about 1 GHz·km in multimode graded index fibers have been achieved by the use of accurate refractive index profiles. The bandwidths in low-mode and single-mode fibers are, of course, substantially higher.

The fact that "communications grade" fibers can now be routinely produced by many different methods has led to a re-direction of emphasis towards the development of economical production methods. However, we intend to completely ignore such technological aspects, and will instead concentrate on the essential features (some of them historical) of the most important processes.

Let us first consider the loss mechanisms, absorption and scattering, that exist in fibers. The fundamental limit on absorption is, of course, the intrinsic absorption in the material, while additional absorption occurs due to ionic impurities (Fig.4.1), as well as atomic defects or colour centers [4.2,3]. The fundamental limit on scattering loss is again intrinsic to the material, while additional sources are possible inhomogeneities in the glass, and longitudinal variations in the refractive index profile. Hence, it is clear that, in general, the materials that are used for fiber production should be such as to allow easy purification and homogeneous (defect-free) formation. Moreover, in view of the potentially large quantities required for communications, the raw materials should be abundant.

These requirements are fulfilled by a number of glasses such as the high silica content, the soda-lime, and the lead-silicate types, but the first of these is probably the best because of the low intrinsic absorption of fused silica. (The intrinsic absorption being difficult to measure, this conjecture is actually based on the lowest attenuations achieved to date). Unfortunately, fused silica has a low refractive index, and it is more difficult to use as the fiber core material, because lower refractive index glasses are not easy to prepare. As will be seen later, this problem is by-passed by using doped silica for the core, and pure silica for the cladding.

As we saw in Chapter 3, the simplest fiber consists of a homogeneous core surrounded by a thick homogeneous cladding. Such step-index fibers were also the first types to be made, by the so-called rod-in-tube and double-crucible methods [4.4]. The former method is almost self-explanatory, and simply involves the preparation of a "preform" of a higher-index rod in a lower-index tube. The fiber is subsequently obtained by uniformly heating and pulling the end of the preform, as schematically illustrated in Fig.4.2. Assuming that surface tension ensures fiber circularity, it is then an elementary exercise to show that

$$\left(a_1/a_2\right)^2 = \ell_2/\ell_1 \quad , \tag{4.1}$$

where a_1 and a_2 are the radii of the preform and the fiber, respectively, while ℓ_1 and ℓ_2 are the corresponding lengths. We deduce from (4.1) that fiber dimensions will have the same ratio as that of the preform. It follows that the size of the fiber can be simply controlled by the speed at which the fiber is drawn.

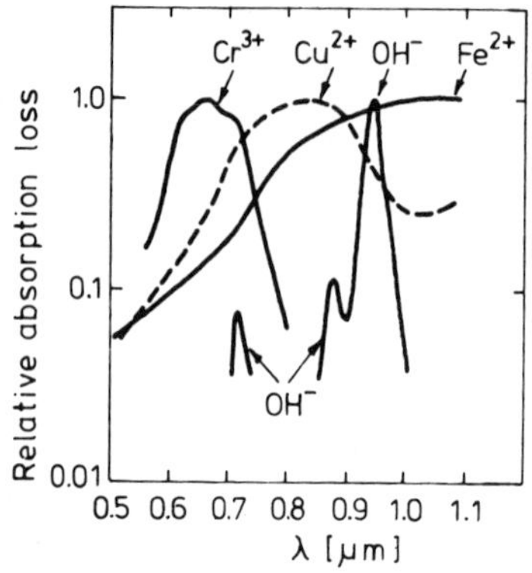

Fig. 4.1. Relative absorption loss versus wavelength for certain ions in glass [4.2]

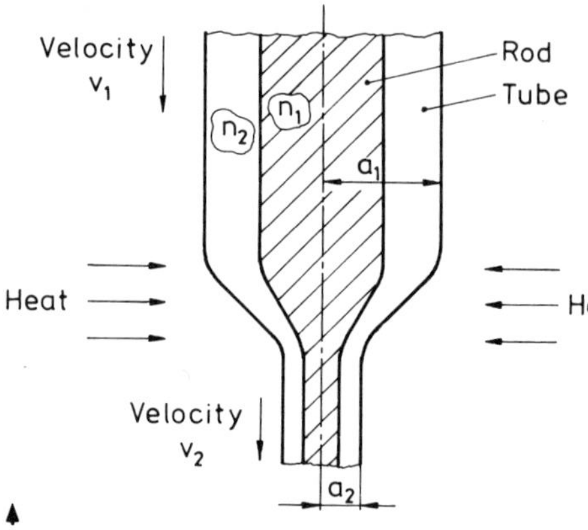

Fig. 4.2. Simplified diagram of the "rod-in-tube" technique

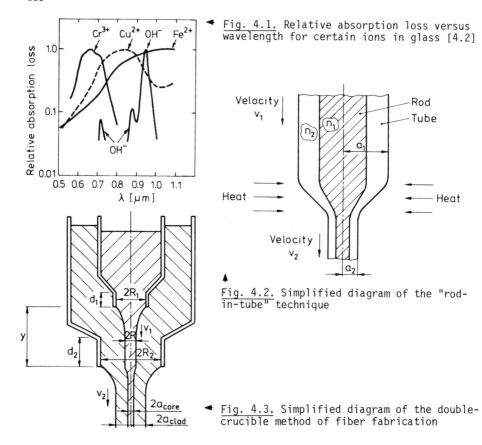

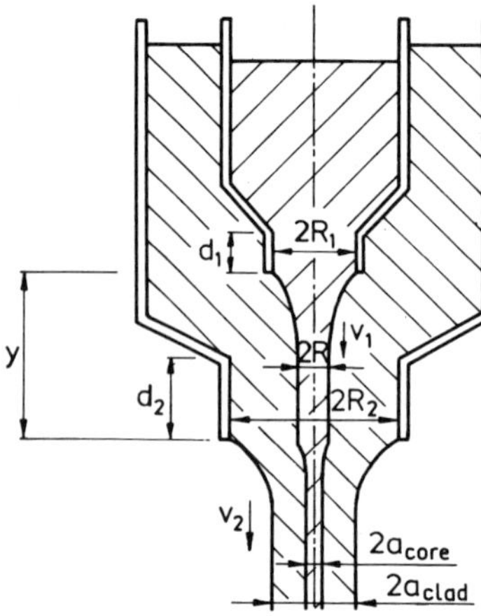

Fig. 4.3. Simplified diagram of the double-crucible method of fiber fabrication

The double-crucible method is somewhat similar, but now, instead of a preform, we have two concentric crucibles (Fig.4.3), the inner one containing molten glass of higher refractive index, and the outer one of lower index. Usually, the melt is formed using glass rods of high purity [4.5]. The crucibles are aligned vertically with the inner one being a few centimetres higher than the outer one. With this set-up, and using the standard equations of Poiseuille flow, it can be easily shown that the ratio of core to cladding radii is given by [4.5]

$$\left(a_{core}/a_{clad}\right)^2 = \frac{P_1\eta_2R_1^4d_2}{P_2\eta_1R_2^4d_1} \ , \tag{4.2}$$

where the subscripts "1" and "2" refer to the inner and outer crucibles, respectively. Also, η is the viscosity, P the pressure difference across the nozzle, R the radius of the nozzle, and d its length. Obviously, the fiber dimensions are now determined in a more complex manner. In particular, the ratio of the two viscosities may vary as a function of temperature (implying the necessity for temperature control), while the two pressures may vary during the pulling operation if the input flow of molten glass is not continuous. Moreover, the meniscus region (Fig.4.3) may have to be protected from the environment, in order to preserve circular symmetry. In principle, this is true for all methods that require the fiber to be drawn.

The double-crucible method can also be used for the fabrication of graded index fibers [4.6], by allowing ionic exchange (by diffusion) between core and cladding glasses. In the method described by KOIZUMI et al. [4.6], the inner crucible contains borosilicate glass with Tl ions, while the outer one with Na ions. The former have small ionic radii and large polarizability, so that the exchange of Na and Tl ions can easily produce the required refractive index difference. We saw in Sect.2.3 that the polarizability of a material directly affects its refractive index, so that loss of Tl ions will decrease it. This loss is obviously largest near the interface between the glasses, so that the decrease in the refractive index will also be the greater there. Using the symbols marked in Fig.4.3, it is easily seen that

$$v_1 = \frac{a_{core}^2 v_2}{R^2} \quad , \tag{4.3}$$

and that

$$Y = v_1 t \quad . \tag{4.4}$$

Moreover, as a measure of the ion-exchange, we can define a parameter

$$K = Dt/R^2 \quad , \tag{4.5}$$

where D is the diffusion constant. Combining (4.3-5), we obtain

$$K = (YD)/\left(a_{core}^2 v_2\right) \quad . \tag{4.6}$$

Numerical evaluation of the diffusion equation shows that near-parabolic profiles can be obtained for K ranging from about 0.01 to 0.1 [4.7]. Thus, by adjusting the parameters in (4.6), a near-parabolic profile of given radius can be obtained. Obviously, ions other than the ones considered here can also be used (see e.g. [4.5]). However, the choice of ions will affect the diffusion constant, which will in turn determine the velocity at which the required fiber can be pulled. We should perhaps mention that, in practice, the obtained profiles are more abrupt near the cladding than those predicted by the computation in [4.7]. The main reason for this is probably that certain effects have been ignored. For example, the diffusion constant has been assumed identical for core and cladding glasses (unlikely), and diffusion is assumed to occur in a region of laminar rather than Poiseuille flow.

The important advantages of the double-crucible method are: 1) the ease with which geometries can be altered, 2) the ease with which glass compositions can be changed, and 3) the potential for extending the process into a continuous one. Disadvantages are: 1) poor profile control, 2) risk of impurity contamination, and 3) the inherent problem of crucibles that are neither attacked by molten glass, nor themselves contaminate it. The availability of compound glasses of consistent quality may also be a problem. However, presumably due to the potential for continuous production, the method has its supporters, and further advances have been made [4.8].

The process that yields the lowest attenuation figures and the most consistent performance is known as the *Chemical Vapor Deposition* (CVD), or the *Inside Vapor Phase Oxidation* (IVPO) technique. The essence of the method is depicted in Fig.4.4. A fused quartz tube is rotated in a lathe-like machine, and is heated to a temperature of about 1600° C by a multiburner torch that continuously sweeps the length of the tube. Oxygen is bubbled through the required reactant, and the resultant mixture is injected into the tube, where it oxidizes and is deposited on the tube wall as "soot". Subsequent sintering yields a transparent glass layer. The chemical reactions may be written down as follows

$$SiCl_4 + O_2 \rightarrow SiO_2 + 2Cl_2 \qquad (a)$$

$$GeCl_4 + O_2 \rightarrow GeO_2 + 2Cl_2 \qquad (b)$$

$$4POCl_3 + 3O_2 \rightarrow 2P_2O_5 + 6Cl_2 \qquad (c)$$

$$4BCl_3 + 3O_2 \rightarrow 2B_2O_3 + 6Cl_2 \quad . \qquad (d)$$

The reaction in (a) gives layers of pure silica, while the other reactions can be used to dope the silica to increase or decrease its refractive index. Reactions (b) and (c) cause an increase while (d) a decrease. Thus, after deposition of the core-cladding boundary layers, subsequent layers are formed by increasing the amount of oxygen bubbled through, for example, the germanium tetrachloride. Of course, various admixtures of reactants are also possible. For example, one may wish to dope with a mixture of P_2O_5 and GeO_2, in order

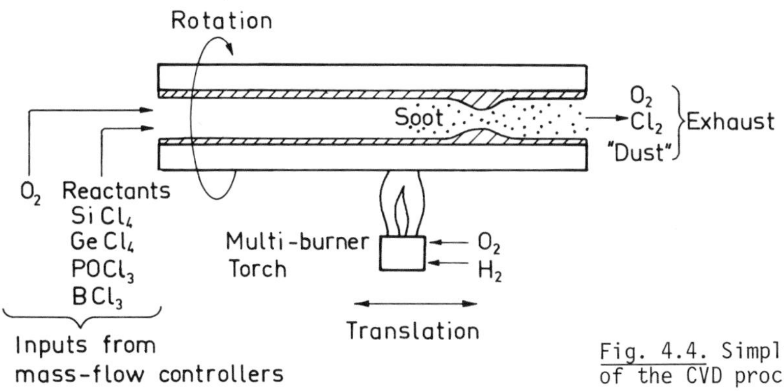

Fig. 4.4. Simplified diagram of the CVD process

to produce specific characteristics [4.9]. In any case, after the required refractive index layers have been deposited, the temperature of the tube is raised well beyond the softening temperature and the tube collapses into a "solid" preform. This preform is then drawn into a fiber, in much the same way as in the rod-in-tube technique.

A major advantage of the CVD method is the high degree of purity that can be achieved by exploiting the enclosed nature of the deposition process. Moreover, a broad range of glass compositions can be used, because the center hole can be closed while the tube is still hot. This eliminates any free surfaces at which fracture-inducing tensile stresses can build up during the cooling process. The main disadvantages are the relatively short length of the preform, and the finite number of layers that can be practically deposited (typically 50-60). As a result, fibers made by this technique often have troublesome ripple in their refractive index profiles.

This difficulty is reduced in the *Plasma-Induced Chemical Vapor Deposition* (PCVD) process developed at PHILIPS [4.10]. Using this process, several thousand layers can be deposited. The small amount of ripple is further smoothed by diffusion during the various heat treatment stages. This method is depicted in Fig.4.5, from which it can be seen that the silica tube is preheated in a stationary furnace (to about 1000° C). Under the stimulus of the plasma; the gases react heterogeneously at the tube wall to directly produce a solid layer of glass. The plasma sweeps the tube very rapidly (~ 8 cm/s), continuously depositing more layers.

Another method that allows the growth of a relatively large number of layers (~ 200) is the so-called *Outer Vapor Phase Oxidation* (OVPO) technique [4.11], shown schematically in Fig.4.6. The raw materials necessary for glass formation are injected via a burner onto a "bait" rod, where they are deposited

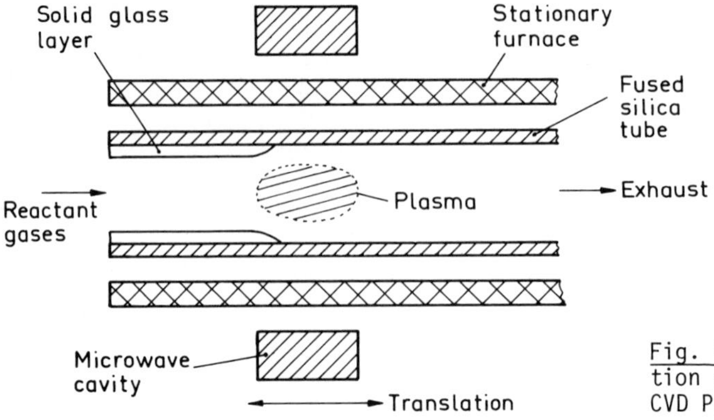

Fig. 4.5. Schematic depiction of the plasma-induced CVD Process [4.10]

in soot form, as shown in Fig.4.6a. The rod is simultaneously under rotation and translation, and as in the CVD method, layers of varying composition can be built up. After removal of the bait rod, the soot preform is finally sintered in a furnace before being drawn into a fiber. These two operations are illustrated in Figs.4.6b and c, respectively. Note also that the central hole closes up during the drawing process.

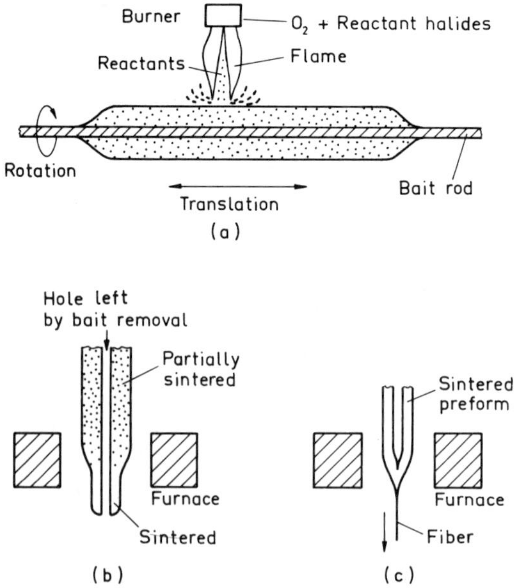

Fig. 4.6. The Outer Vapor Phase Oxidation (OVPO) process [4.11]. (a) Soot deposition (b) Preform sintering operation (c) Fiber drawing operation

The main advantages of the OVPO technique are the precise control of the refractive index profile, and the ease with which relatively large preforms can be fabricated. For example, more than 10 km of 125 μm (outer diameter) fibers can be readily produced [4.11], while the bandwidth of production fibers is in excess of 1.2 GHz · km [4.12]. Amongst the disadvantages are: 1) the existence of the central hole after bait removal, and 2) the introduction of hydroxyl impurity by the flame combustion. The former can result in expansion stresses and breakages of sintered preforms of high numerical aperture, while the latter affects the attenuation, particularly at the longer wavelengths. However, these problems can be largely avoided by the use of the proper technology. For instance, the hydroxyl contribution can be dramatically reduced by using gaseous chlorine drying in the sintering process, whereas stress reduction can be achieved by the use of stress-balancing concepts [4.11].

Another technique for fiber fabrication that completely avoids the central hole, but uses outside deposition, is the so-called *Vapor-Phase Axial Deposition* (VAD) method, depicted in Fig.4.7. Gaseous raw materials, such as $SiCl_4$, $GeCl_4$, and PCl_3, are fed into an oxy-hydrogen burner. The resulting stream of fine glass particles (due to flame hydrolysis) is directed towards one end of the starting rod, where it is deposited in porous form. A second burner is used to blow boron doped particles of lower refractive index. By changing the proportions of the various materials, and by adjusting the rates at which the particles are deposited, it becomes possible to create a radially-graded profile. If necessary, more than two burners can also be used. In any case, the starting rod is continuously rotated, and is simultaneously moved upwards at a speed consistent with the growth rate of the preform. During its upward motion, the porous preform passes through a ring-like furnace where it is sintered. If required, hydroxyl (OH) ions can be removed by heat treatment in a thional chloride ($SOCl_2$) atmosphere [4.13], usually before the sintering operation. The main advantages of the VAD method are: 1) the potential for the fabrication of large preforms, 2) complete avoidance of the central hole found in the OVPO preform, and 3) the attainment of low-attenuation fibers when enclosed deposition is used. Perhaps the main disadvantage is the difficulty of accurate profile control.

In summary, we have discussed the essential features of some of the fiber fabrication techniques that are in use. Each method has its own merits in terms of fiber performance and ease of production, and should be carefully considered. Moreover, although each method has been separately presented, a certain amount of "intermixing" can also occur. For example, in the pre-

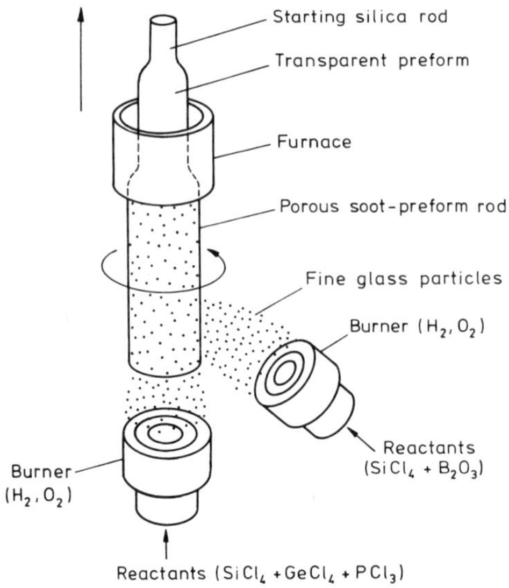

Fig. 4.7. The Vapor-Phase Axial Deposition (VAD) Method [4.13]

paration of single mode fibers, the core section may be grown as a separate preform that is then inserted into a silica tube, in much the same way as in the "rod-in-tube" method. As another example, glass rods used in the double-crucible method may be formed using a technique similar to the OVPO process, in order to obtain a high level of purity.

A feature that is common to all the described methods is the necessity to "draw" the fiber. This requires furnaces capable of *evenly* heating the preform, and rotating drums that take it up. Often, the speed of the drum is controlled by monitoring fiber thickness, and usually the fiber receives a primary coat (e.g. some type of plastic) before take-up. However, all of these aspects are essentially technological, and have been neglected. In fact, the technological problems represent an art in themselves (see e.g. [4.14]), and must inevitably remain outside the scope of this introductory text.

4.2 Optical Sources

In broad-band fiber systems, the requirement of low modal dispersion necessarily constrains us to fibers with small cores and low numerical apertures. From the viewpoint of source-to-fiber coupling efficiency, the implication is that we should also restrict ourselves to sources that are small and have low beam divergences. If we further include such practical considerations as low cost, large bandwidth, and ease of modulation, our choices are narrowed down to semiconductor lasers, fiber lasers, and light-emitting diodes (LEDs).

We saw in Sect.2.1 that LEDs and semiconductor lasers are very similar physically in that both depend on electron-hole pair recombinations for light generation. It was also mentioned that, in principle, population inversion could be achieved for a sufficiently high density of injected carriers. However, in such "homo-junctions" the required drive current tends to be impractically high. This problem is normally avoided by the use of "hetero-junctions" [4.14-16], which result whenever a junction is formed between semiconductors with different bandgaps. Although the use of such hetero-junction is not necessary for LED operation, their use considerably improves the efficiency and bandwidth of the diodes. Thus, theoretically, we can justifiably consider hetero-junction LEDs to be the special case of hetero-junction lasers. In this section we will therefore restrict ourselves to a brief discussion of semiconductor and fiber lasers, and refer the reader to [4.17-20] for more detailed treatments of LEDs as well as semiconductor lasers.

In optical fiber communications, the most widely used laser diode structure consists of a double hetero-junction (DH) between GaAs and AlGaAs. Amongst the many variants that are used, the simplest to understand consists of a lightly doped GaAs layer sandwiched between two p and n doped $Al_xGa_{1-x}As$ layers (Fig.4.8). The resultant energy diagram is also shown in Fig.4.8, and we see that the GaAs region constitutes a potential well that forces the injected carriers to "dwell" in this region. We can immediately reach two conclusions: 1) that the probability of recombination in the GaAs region must now be higher than in a homo-junction, and 2) that a correspondingly smaller current should be necessary to maintain population inversion. From the viewpoint of device characteristics, the implication is an increase in efficiency,

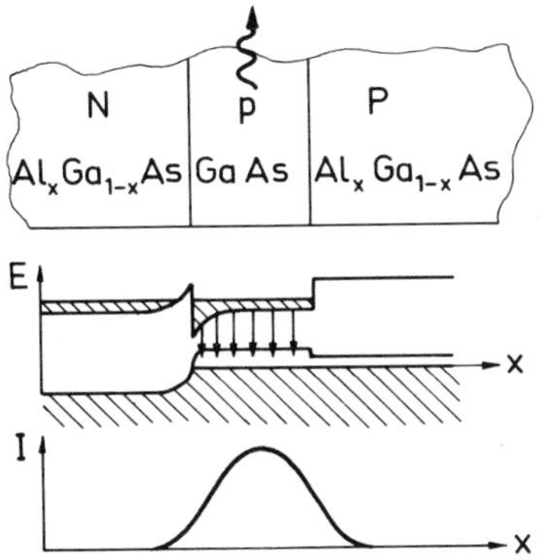

Fig. 4.8. Schematic representation of a DH laser (top), its energy diagram (middle), and an idealized near-field profile

and a decrease in the threshold current at which laser operation is initiated. In any case, we see that the GaAs region is our active material from which the optical oscillator should be constructed.

The next requirement is to obtain positive feed-back by, for example, cleaving the ends of the GaAs region. The reflectivity of the GaAs-air interface is approximately 30%, so that a useful resonator can be obtained in this way. Moreover, because AlGaAs has a lower refractive index than GaAs, we also have a planar waveguide between the AlGaAs layers (see Sect.3.3 and 4). Since, in our example, the two sandwiching layers are equally doped and thus have equal refractive indices, we are dealing with a symmetrical waveguide. However, it must be remembered that the analyses of Sections 3.3 and 4 are not directly applicable here, because we now have a "boxed" resonator, whereas in our simplified analyses, an infinitely long waveguide was assumed. Nevertheless, we can at least qualitatively conclude that our diode waveguide will have certain allowed modes into which the generated light will be coupled. In addition, we expect the modal fields to extend outside the GaAs region. Fortunately, with the normally used Al concentrations, AlGaAs is relatively transparent at the wavelength in question ($\sim$900 nm), so that little additional loss is incurred. Another conclusion we can reach is that, by shrinking the size of the GaAs layer, we should be able to obtain a single transverse mode. The condition for this can be shown to be [4.21]

$$0.022\left(\frac{\lambda}{d}\right)^2 < x < 0.07\left(\frac{\lambda}{d}\right)^2 \quad , \tag{4.7}$$

where d is the thickness of the GaAs layer, λ is the generated wavelength, and x is the Al mole fraction.

We see that both charge carriers and photons can be confined in the transverse direction (perpendicular to the junction plane), but have not considered confinement in the direction parallel to the plane of the junction. A widely used method for achieving this is to use a stripe contact such as the one illustrated in Fig.4.9. In this example, the stripe is defined by an etched well in a reverse biased p-n junction, so that conduction only occurs under the stripe region. Another commonly used technique for carrier confinement is proton bombardment [4.22], which increases the resistivity of that part of the semiconductor that is not under the stripe. However, care must normally be taken to prevent proton penetration into the active area in order to avoid undesirable changes in the optical properties. A host of other stripe geometry laser configurations are also used, most of them summarized in

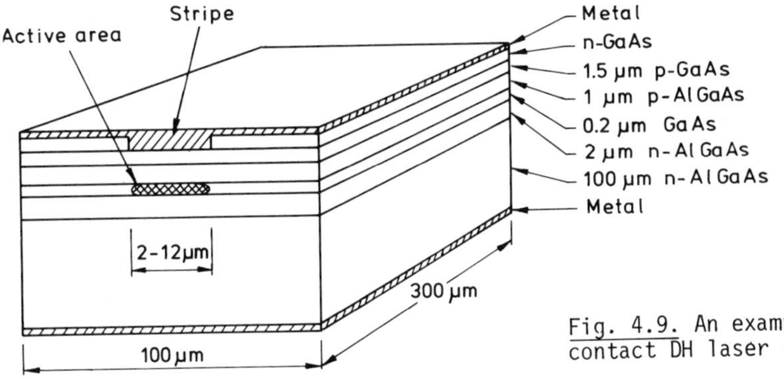

Fig. 4.9. An example of a stripe-contact DH laser (not to scale)

[4.18,19]. One reason for the large number of configurations is the existence of non-linearities, called "kinks", in the output characteristics of many stripe contact lasers.

We can now see that confinement has been obtained in all directions: longitudinal confinement by the cleaved facets, transverse confinement by the hetero-junctions, and lateral confinement by the contact stripe. In fact, the latter not only confines injected carriers but also waveguides the optical field via the slight increase in refractive index produced by the combined effects of 1) local heating which increases the real part of the refractive index, 2) increasing gain which increases the imaginary part, and 3) the existence of free carriers which decrease the real part. In view of the foregoing, it is easy to understand that, even if we would only have one lateral mode when operating near threshold, sufficiently far away from threshold the refractive index would reach a value where the cavity would have two allowed lateral modes. Such mode changes are often accompanied by the "kinks" mentioned earlier. As a result, we see that the stripe width may play a critical part in the avoidance of nonlinearity. The alternative, more complex, approach would be to build in more definite refractive index steps in order to achieve lateral confinement.

Let us next consider some basic external properties of DH stripe lasers. The most important characteristics that, incidently, also provide useful information about the laser's internal behaviour, are: 1) the relationship between the output optical power and the input drive current, 2) the spectral behaviour of the output, and 3) the near and far-field power distributions.

The power-current characteristics can be qualitatively understood in terms of rather simple concepts. Consider, for example, the idealized model of the stripe laser shown in Fig.4.10a. The injected current I is distributed under

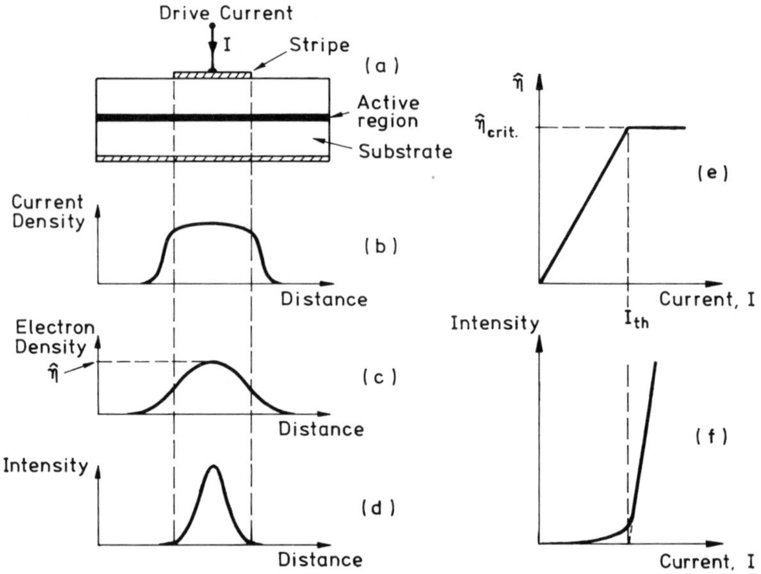

Fig. 4.10a-f. Behaviour of an idealized DH stripe laser: (a) Simplified structure. (b) Current density profile in active region. (c) Electron density profile. (d) Resultant output intensity. (e) Maximum electron density against drive current. (f) Output intensity against drive current

the stripe and spreads out because of the non-ideal electric field that arises from the finite resistivity of the surrounding semiconductor. The corresponding current density is shown in Fig.4.10b. The electron density profile in Fig.4.10c does not follow the current density exactly, because of diffusion and stimulated recombinations. Moreover, a certain threshold value of the electron density is required before the net gain in the cavity can exceed the net loss. This results in the narrower intensity profile in Fig. 4.10d. If we now assume that the shapes of these profiles remain independent of current, we can also plot the behaviour of peak electron density as a function of input current, as in Fig.4.10e. We see that the electron density saturates at the threshold current I_{th}, and beyond this point a more or less constant fraction of all the injected carriers contribute towards stimulated emission. As a result, we obtain the ramp-like rise of the intensity in Fig. 4.10f.

The second important characteristic mentioned above is the spectral distribution of the output, and can be easily understood on the basis of our discussion of Fabry-Perot resonators in Sect.2.1. In fact, present day technology allows the fabrication of single-longitudinal-mode diodes, and a number of devices are commercially available. The output spectrum of one such device is shown in Fig.4.11, from which it can be seen that, although some power still exists in other modes, the fundamental mode is strongly dominant.

The third set of aspects mentioned above are the near- and far-field power distributions. These are useful for studying the modal distributions in the laser cavity, but are also particularly useful for estimating the coupling efficiency with which power can be excited into a fiber. Two directions of interest can obviously be specified, since the active region is defined by a rectangle of about $10 \times 0.2\mu m$. Thus, in the direction perpendicular to the junction, we expect a larger divergence, while parallel to the junction we expect a smaller one, since the "slit" width is more in this direction. From the viewpoint of coupling, the former is usually more important, particularly for multi-mode fibers that can completely accomodate the distribution in the other direction. This can be confirmed from Fig.4.12, where the far field of a high quality commercial diode is depicted.

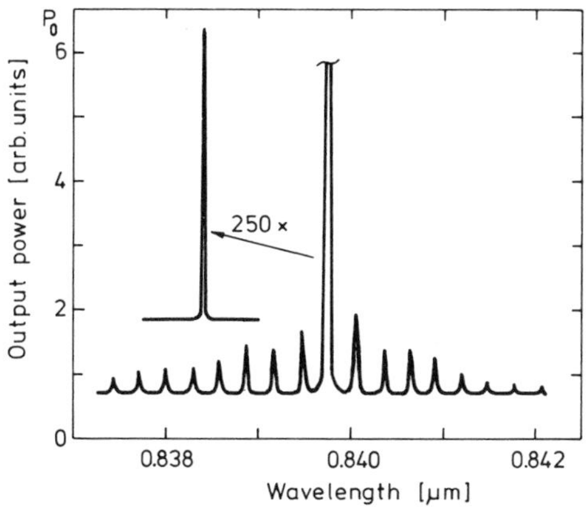

Fig. 4.11. Spectral distribution of a diode operating with a single longitudinal mode

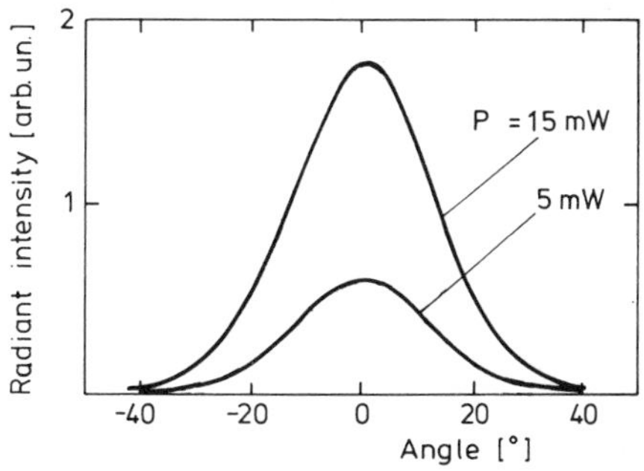

Fig. 4.12. The far-field power distribution of a high quality DH stripe laser perpendicular to the plane of the junction

Another problem to consider is related to the transient behaviour of DH stripe lasers when excited by fast-rising current pulses. Consider, for example, the situation in which an unbiased diode is excited by a current step whose peak value is more than the value of the threshold current. The carrier build-up is initially delayed by the finite life time of the carriers, but eventually the concentration exceeds the threshold level necessary for amplification. The resultant optical field depletes the concentration of carriers, with a consequent decrease in the optical field. Hence, the carrier concentration builds up once again and the cycle is repeated until a steady state is reached. The observed optical output thus contains *relaxation oscillations* whose frequency can range from several hundred megahertz to a few gigahertz. This type of interaction between photons and carriers is obviously undesirable for high speed communication systems.

The turn-on delay could, in fact, be avoided by biasing the diode slightly above threshold. Unfortunately, under such conditions many lasers exhibit the troublesome phenomenon of self-pulsation, usually at frequencies that are noticeably less than the corresponding relaxation frequencies. This problem can be eliminated, without a significant increase in the turn-on delay, by biasing the laser just below threshold. However, in this case another difficulty, the so-called pattern effect, has to be contended with. This is the effect whereby the diode exhibits a "memory" for the preceding (near-by) pulse which produces an increase in the carrier density. Hence, if the next pulse appears before depletion of the excess concentration, the output optical pulse has a larger amplitude than that of the preceding pulse. Several electronic means have been proposed for eliminating this effect [4.23,24] at the expense of increased modulation complexity. One method [4.23] is to logically insert a compensation pulse just before all modulation pulses that are preceded by an empty time slot. The level of the compensation pulse is arranged to be insufficient for light generation, but sufficient for raising the carrier concentration to the level that would have been produced by a legitimate pulse. Thus, all amplitudes remain constant. Alternatively, the excess carrier concentration, due to each modulation pulse, could be extracted by injecting a negative pulse immediately after every modulation pulse [4.24].

Even after all these complex measures for eliminating turn-on delay, we are still left with the relaxation oscillations, which can only be reduced by introducing a damping factor in the resonance. One effective way is to ensure a relatively large contribution from the spontaneous emission process [4.25 and Ref.4.18, Chap.7]. For example, laser diodes with a "soft" current-power characteristic behave like LEDs up to the threshold current, so that they provide a relatively large photon flux. In diodes of this type, relaxation oscillations are rarely a problem. Another approach is to inject coherent light into a single oscillating mode, in which case powerful damping can be achieved without multimode operation.

Although, the relaxation effect is generally a draw-back for high speed communications, we should note that it can also be used to advantage for the generation of short optical pulses, by injecting a suitably tailored current pulse into the diode. The width and amplitude of the energizing pulse obviously varies from diode to diode, but the basic pricniple is to allow the first optical "over-shoot" to destroy the population inversion.

Our foregoing discussions indicate that the main performance limitations associated with end-cut (Fabry-Perot) DH stripe lasers are: 1) the relatively low reflectivity of the end mirrors, 2) the difficulty of achieving and maintaining modal "purity", and 3) the existence of transient relaxation effects. These reasons lead us to the consideration of thin film lasers, in which, at least conceptually, it should be possible to avoid relaxation oscillations by "integrating" two lasers on the same substrate. Moreover, with the use of some corrugated waveguide type of feedback, the reflectivity of the "resonator" can also be improved substantially. In addition, such a structure is more convenient for ensuring single-mode operation.

Thin-film lasers of the above type fall into two main categories, depending on the method of feed-back. One is based on the Distributed Bragg Reflector (DBR), while the other uses Distributed Feed-Back (DFB). The DBR structure is illsutrated in Fig.4.13a, and is somewhat similar to the Fabry-Perot resonator in that the reflecting regions are on either end of the active layer. On the other hand, the DFB resonator bears no resemblance to the Fabry-Perot type. In both DBR and DFB structures the actual process of reflection is based on the Bragg diffraction of the incident wave at the periodic interface. The periodicity may be due to variations in the refractive index, and/or the optical gain of the active material. Alternatively, the periodicity may be introduced by corrugations, e.g. by etching or ion milling, because this is effectively equivalent to the introduction of refractive index variations.

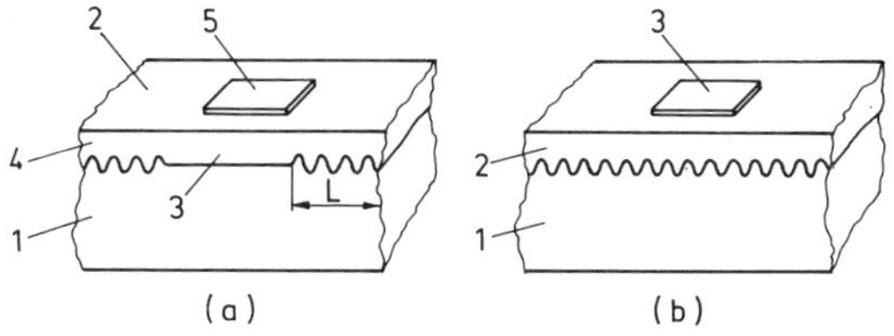

Fig. 4.13. Schematic illustration of the two main types of thin-film lasers: (a) The DBR type with separated reflecting regions. (b) The DFB type with a single Bragg resonator

Some of the earliest investigations on the use of Bragg diffraction in thin-film lasers were carried out by KOGELNIK and SHANK [4.26,27] using thin dye films. However, these were soon followed by the observation of laser action in optically pumped GaAs-AlGaAs at low temperature [4.28-31]. Injection pumped laser action was first observed at low temperatures [4.32], followed by pulsed operation at room temperatures [4.33]. After this rapid development, CW operation at room temperature was eventually attained by NAKAMURA et al. [4.34]. Today, the interest is mainly centered on the improvement of semiconductor-based thin-film lasers.

Physically, the operation of the Bragg resonator can be simply understood by considering each surface perturbation to be the source of partial backward reflection. Thus, if the periodicity is such as to produce a propagation delay that corresponds to the condition for constructive interference, rather powerful feedback can be obtained. In its simplest form the required periodicity can be estimated from the formula [4.35]

$$\beta_d = \beta_i - \frac{2\pi m}{\Lambda} , \qquad (4.8)$$

where β_i is the longitudinal propagation constant of an allowed incident mode, β_d is the propagation constant of an allowed diffracted mode, Λ is the period of the corrugations, and m is an integer. With Λ equal to half the wavelength, i.e. $2\Lambda = \lambda = 2\pi/\beta_i$, and m = 1, we obtain $\beta_d = -\beta$, and the diffracted wave of the first order will be a reflection of the incident wave.

An important feature of Bragg resonators which emerges in a more detailed analysis [4.36,37] is that the "reflection" coefficient is rather insensitive to small amounts of "de-tuning". Hence, small deviations from the optimum structure do not necessarily hinder operation. The survey of experimental results in [4.36,37] also indicate the potential superiority of DFB and DBR lasers over end-cut diodes. Moreover, as mentioned earlier, the lack of end mirrors encourages us to anticipate dual integrated lasers with one providing coherent injection for the other, and thus reducing relaxation oscillations. Of course, the assumption is that the locking range of the dual system is larger than any possible de-tuning. For the host of technological difficulties that are associated with thin-film lasers of the discussed type, as well as analytical details, we refer the reader to the cited works.

Finally, let us briefly consider the structure of neodymium-based fiber lasers. In principle, these lasers are very similar to their bulk counterparts, with the bulk rod replaced by a thin fiber. The interest in such devices is due to at least two reasons: 1) the dimensions of the fiber rod can be made compatible with multimode fibers in order to achieve high coupling efficiency, and 2) use of the rod-like material makes it easier to achieve high reliability, modal purity, and a narrow spectral width. In fact, the weakest link, from the viewpoint of reliability, is the life time of the pumping source. The use of long-lived LEDs for pumping alleviates even this difficulty. Another advantage of the Nd laser is that its important emission wavelengths lie in the range 1.05-1.35 μm where the attenuation and material dispersion of present day silica fibers is low. Perhaps the major drawback with Nd is that its upper laser level has a long fluorescence lifetime, on the order of 0.1 ms as compared to about 1 ns in GaAlAs injection lasers. Hence, direct high speed modulation (via the pump source) is not possible, and an external modulator becomes necessary.

Although a number of host materials could be used, the best fiber lasers reported to date are based on yttrium-aluminium-garnet (YAG) as the host [4.38-40]. This crystalline material is rather useful because of its relatively low photo-elastic constants, as well as good optical, mechanical, and thermal properties. The composite structure forms a four-level system (Sect. 2.1) with the strongest pumping band located at 0.81 and 0.75 μm. We see that the use of specially tailored LEDs would allow us to exploit these bands.

In contrast to bulk Nd:YAG lasers, optical pumping via the curved surface of the fiber is not particularly attractive because of the extremely small dimensions. The alternative is to inject the radiation from a LED into one end of the fiber via a mirror which is transparent to the LED spectrum but not to the emission spectrum of Nd:YAG. A structure based on this approach is depicted in Fig.4.14. The fiber rod (1) is a short section ($L \sim 5$ mm, $\phi \sim$ μm) of a longer fiber drawn from a crystal of Nd:YAG. This rod is covered by a cladding of glass (2); and placed in a silica tube (3) that provides rigidity for the fiber and supports the output mirror (5), and the input mirror (4). The mirrors are bonded onto the silica tube using epoxy resin, and the resonator is aligned with the aid of a Kr laser during the "potting" life of the resin. STONE and BURRUS [4.40] report a CW output of 1 mW at 1.064 μm for a structure of this type. The associated intensity distribution is of the form shown in Fig.4.15. The efficiency with which the pump radiation is absorbed can be estimated from Fig.4.16, in which the dashed curve shows the spectrum of the LED, while the solid curve shows the remaining power after absorption by 10 mm of fiber. We see that, in this particular case, about 50% of the LED's power is absorbed.

In concluding this section, we would like to remind the reader why modal purity and stability are important factors. As will be seen in Sect.6.1, the inter-repeater spacing in a practical system is normally calculated on the basis of a certain constant power at the input of the communication fiber. Any unpredictable changes due to aging etc. are normally accounted for by

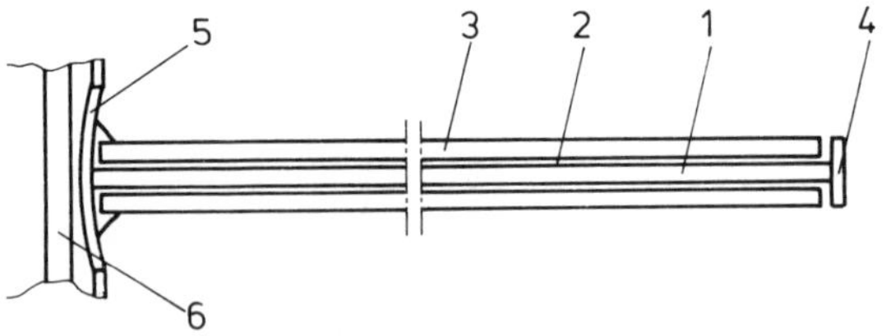

Fig. 4.14. Basic structure of an end-pumped fiber laser with the numbered parts described in the text [4.40]

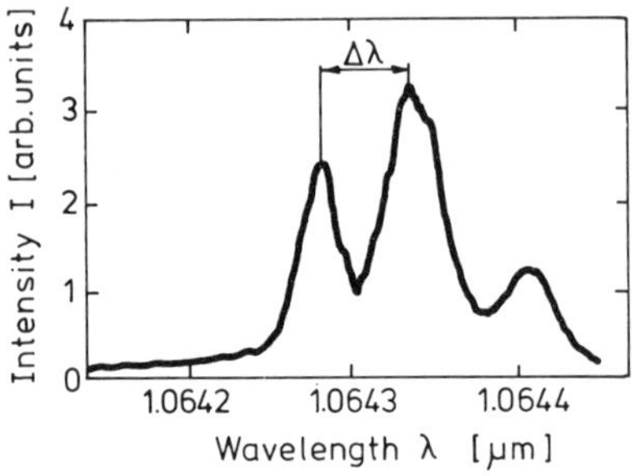

Fig. 4.15. Output spectral distribution of the fiber laser of Fig. 4.14 [4.40]

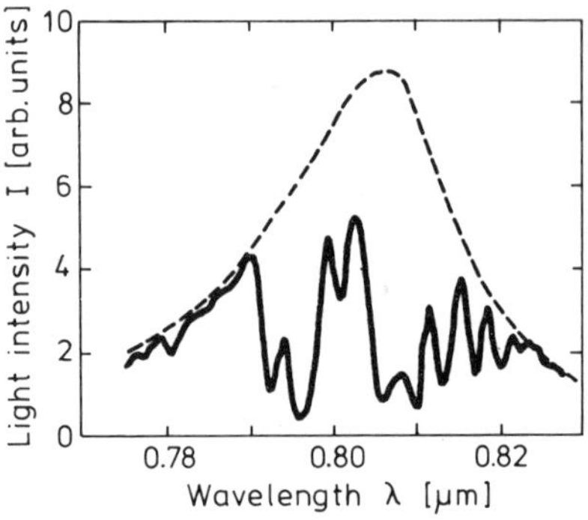

Fig. 4.16. The spectral distribution of the pumping radiation at the input of the fiber laser (---) and after 10 mm (——) [4.40]

an impairment allowance. Temporal variations in the output mode structure can lead to fluctuations in the coupling efficiency, and hence to changes in the input power. Therefore, if the source is not stable, a much larger impairment allowance must be made to account for the worst case situation.

4.3 Thin-Film Modulators

We saw in Sect.2.2 that a large number of physical effects could potentially be used for the modulation of optical signals. In this section we will restrict ourselves to a review of some of the most promising thin-film structures that exploit the electro-optical effect. The main justifications for this restriction are i) that a large variety of solutions based on the electro-optical effect have been reported, and ii) that large modulation bandwidths can be achieved using these solutions.

The basic features that should be considered in the design of such thin-film electro-optical modulators, can be broken down into two distinct groups:

i) the preparation of an electro-optical waveguiding layer, and ii) the selection of an appropriate electrode configuration as dictated by the "concept" of the device. Here, we will ignore all questions related to the difficulty of coupling the modulator to a fiber at one end and a source at the other.

Let us then start from the waveguiding layer and note that a number of materials such as $BaTiO_3$, KTN, CdS, ZdS, ZnO, $LiNbO_3$, and $LiTaO_3$ are available. Of these, the last two have been rather extensively investigated, and a number of techniques for the fabrication of thin films from these materials have emerged. The essential principles of the most significant of these techniques are listed below:

1) *Out-diffusion* [4.41] exploits the high mobility of Li_2O molecules to reduce the ratio of the Li:Nb surface concentrations. The main disadvantages of this method are: (a) that only the extra-ordinary refractive index is different from that of the substrate, and (b) that three-dimensional "strip" waveguides are difficult to produce.

2) *In-diffusion* is the complement of the above in that TiO_2 or Ti is diffused into $LiNbO_3$ or $LiTaO_3$ to produce surface layers of $LiTi_xNb_{1-x}O_3$ or $LiTi_xTa_{1-x}O_3$ [4.42-44]. Another alternative is the in-diffusion of Nb into $LiTaO_3$ to give $LiNb_xTa_{1-x}O_3$ [4.45]. With the in-diffusion technique, both the ordinary and the extra-ordinary refractive index of the surface layers differs from that of the substrate. Moreover, narrow strip widths down to about 5 μm can be fabricated, so that lateral modes can be rather well defined.

3) *Sputtering* of $LiNbO_3$ onto $LiTaO_3$ substrates [4.46] can be used to produce crystalline films whose thickness can be accurately controlled. Moreover, a relatively "sharp" step-index profile can be achieved. Unfortunately, sputtered films presently have somewhat high attenuations: about 8 dB/cm, in contrast to 1 dB/cm for films obtained by diffusion.

4) *Liquid-Phase Epitaxy* (LPE) [4.47,48], or epitaxial growth by melting (EGM) [4.49], can also be used for growing films on e.g. $LiTaO_3$ substrates. The former method yields a refractive index profile which is close to a step, but the loss in such films is about 5 dB/cm. The latter technique provides films of a lower loss ($\sim$1 dB/cm), but it is difficult to achieve step profiles.

5) *Chemical Vapour Deposition* of $LiNbO_3$ films on $LiTaO_3$ substrates [4.50] can also be used, but the absorption loss of such films is presently much higher ($\sim$40 dB/cm) than all the other methods mentioned above.

Let us next briefly survey some of the device structures that have been reported in the literature, in order to familiarize ourselves with the main concepts that are involved, and with the electrode configurations that are used. The simplest and most straightforward of these is depicted in Fig.4.17a and was practically demonstrated by KAMINOW et al. [4.51]. The original structure was obtained by the in-diffusion of Ti into $LiNbO_3$ to yield a three-dimensional waveguide. The control electrodes were obtained by the deposition of metallic stripes on either side of the waveguide, while coupling was achieved using prisms. With structures of these types, useful phase shifts can be obtained for applied voltages of about 0.3 V.

The performance of the above type of modulator can be improved by at least two modifications. The first is to introduce a "ridge" waveguide of the type proposed by UCHIDA [4.52], and illustrated in Fig.4.17b. The advantage of this approach is that the power dissipation can be substantially decreased in comparison to the planar structure. The second solution which still retains plane electrodes is the two-branch interferometer depicted in Fig.4.17c. This type of modulator behaves as a phase-to-amplitude converter [4.53], in the same way as classical interferometers.

In contrast to the above three modulators, an entirely different approach is based on the properties of periodic structures [4.54,55]. This type of configuration is illustrated in Fig.4.17d, from which it can be seen that the "fingered" electrodes will produce a periodic distribution of the applied electric field. The consequent periodicity in the refractive index produces a grating-like structure which deflects the incident optical field by an amount that is determined by the applied electric field. Needless to say, lateral confinement is unnecessary in this case. Another periodic structure is based on the properties of distributed-feed-back (DFB) reflectors [4.56], discussed earlier in Sect.4.2. When such a reflector is positioned between two electrodes, as in Fig.4.17e, the applied voltage produces a change in the Bragg condition, with the result that the reflected and transmitted powers are also correspondingly modified. Unfortunately, technological difficulties hinder the implementation of such structures, and this alluring proposal still awaits its day.

Another useful concept is based on the deflection of light within surface waveguides. One approach uses the resultant change in the angle of refraction at an interface between two media, due to a change in the refractive index of one of the media. If a beam within a film is allowed to cross the boundary between two different regions with slightly different refractive indices, as in Fig.4.17f, then a small change in the refractive index of the second re-

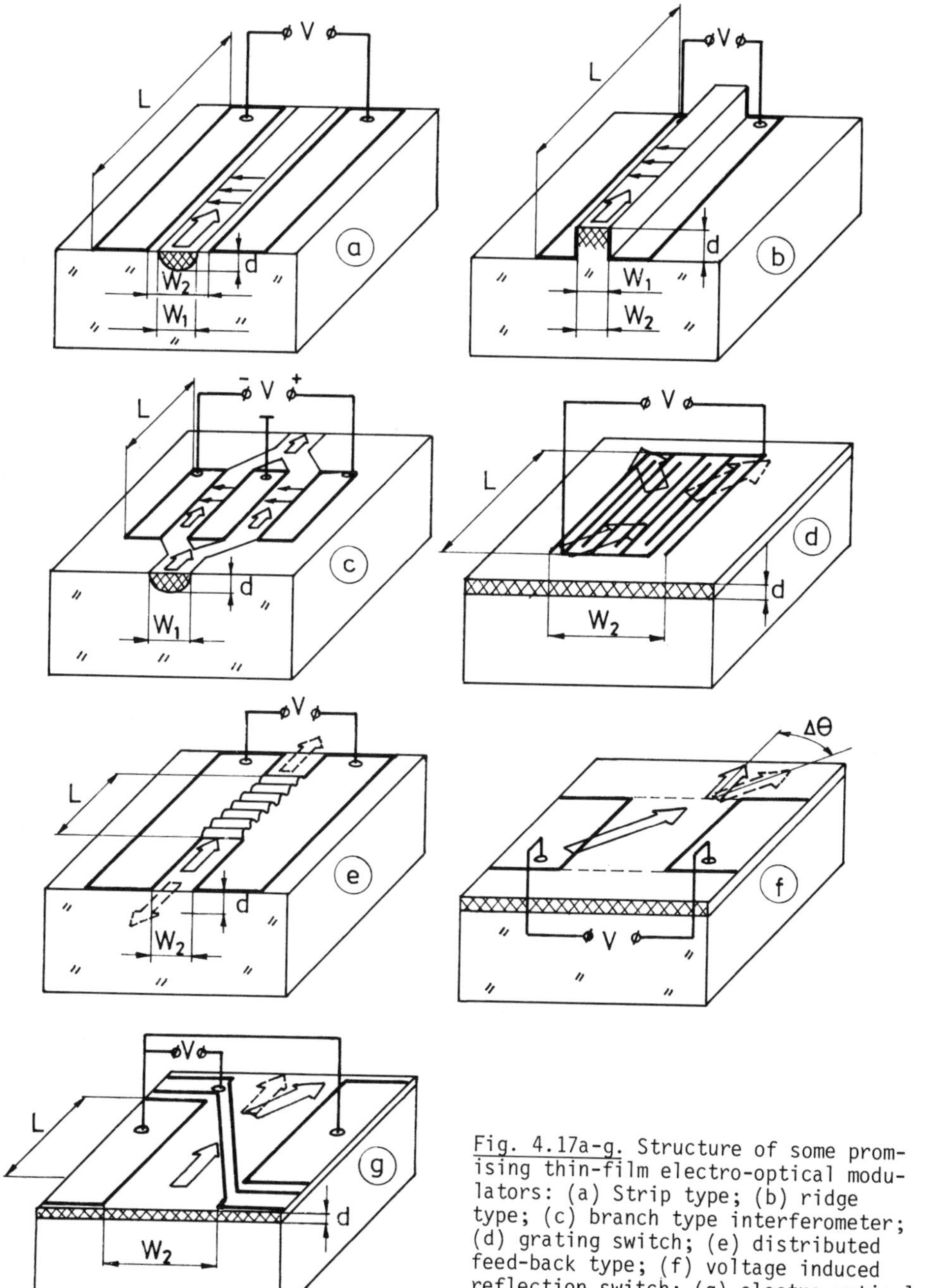

Fig. 4.17a-g. Structure of some promising thin-film electro-optical modulators: (a) Strip type; (b) ridge type; (c) branch type interferometer; (d) grating switch; (e) distributed feed-back type; (f) voltage induced reflection switch; (g) electro-optical prism switch

gion will produce an almost proportional shift in angular position [4.55]. If the Pöckels effect is responsible for the refractive index change, the angular deflection is also proportional to the applied voltage. This idea can also be used for the implementation of the so-called electro-optical-prism switches [4.57,58], as illustrated in Fig.4.17g. The deflector consists of three electrodes above the electro-optic waveguiding film. When voltage is applied to the central electrode and the outer ones are grounded, the applied electric field is non-uniform in the direction of optical propagation. This produces a corresponding gradient in the refractive index, and the beam is deflected as shown in the figure. Here, we should note that although devices based on beam refraction or diffraction are normally called switches, in principle, they can also be used as intensity modulators, provided that the deflection is small so that it only partly shifts the beam from the aperture of the output fiber.

Table 4.1. Characteristics of some electro-optical thin-film modulators described in the text: single asterisk refers to estimated values, two asterisks to 100% modulation, and three asterisks to 80% modulation. V is the applied voltage, α the absorption loss, while the other symbols are defined in Fig.4.17 and in Sect.2.2

Modulator type	λ [μm]	L [mm]	d [μm]	w_1 [μm]	w_2 [μm]	$P/\Delta f$ [mW/MHz]	V [V]	Δf [MHz]	α [dB]	Ref.
Strip waveguide*	0.63	30	< 1	4.6	9.3	1.7×10^{-3}	0.3	530	< 1	4.51
Strip waveguide*	1.05	16.5	6	6	8	-	2	2000	-	4.52
Ridge waveguide*	1.05	16.0	7.5	7.5	7.5	-	1	2000	-	4.52
Bragg reflector*	0.89	1	-	-	5	7.3	17**	430	-	4.56
DFB*	0.89	10	-	-	5	27	46**	780	-	4.56
Grating	0.63	3	~ 1	-	400	5	25***	1000*	-	4.54
Prism	0.63	6	~ 1	-	100	-	15	-	-	4.57
Prism	0.63	-	-	-	-	1	8	-	-	4.58
Interferometer	0.63	3.3	~ 2	6.1	-	-	16	1000	-	4.53

Finally, we would like to draw the reader's attention to Table 4.1 where the main features of the modulators discussed above are summarized. We see that not only are most of the required drive voltages and powers within manageable limits, but the available bandwidths are also quite respectable.

4.4 Photo-Detectors

We saw in Sect.2.3 that PIN and APD diodes are the most significant detectors for optical fiber communications, because they are small and hence compatible to fibers. Also, when based on well understood materials, such as silicon, they can satisfy a number of other needs related to speed, responsivity, noise, and bias voltage. Such diodes are useful for fiber systems operating at the lower transmission "window" around 850 nm, and even at 1.06 μm if a sufficiently wide absorption region is used. However, around the 1.3 μm and 1.55 μm windows, other materials with smaller band-gaps are required. Germanium is useful up to about 1.6 μm, but is not optimum at 1.3 μm, because of its needlessly small bandgap for that wavelength. From this viewpoint, compounds such as InGaAsP and GaAsSb are more attractive, because their band-gap can be tailored for the specific application by simply changing the relative concentration of constituents [4.59,60]. In this way, noise generation due to random thermal excitation can be kept to a minimum by selecting a bandgap that is only slightly less than the photon energy in question. Unfortunately, both in germanium and in III-V alloy devices, many technological difficulties must be overcome before they can offer a performance that is on par with silicon diodes. As such, in the rest of this section, we will ignore detectors for the "longer wavelengths" and restrict our attention to silicon devices. However, since we intend to take an "engineering approach" to detection, restriction to silicon diodes should create no great loss in generality.

Let us first take a brief look at some popular diode structures [4.61-63], as illustrated in Figs.4.18 and 19. Figure 4.18a shows a typical front illuminated PIN diode in which photons reach the absorption region via an antireflection coating and a thin p^+ region of high conductivity [4.63]. Because of the latter, the almost intrinsic ν region experiences a relatively uniform electric field, as indicated by the "flux" lines in Fig.4.18a. Since all other regions are arranged to be thin, most electron-hole pairs are generated in this "drift field" region, so that such devices usually have good rise times ($\sim$1 ns). The absorption efficiency, particularly at longer wavelengths, can be further improved by using a reflective lower contact. The other alternative is to use an interference film reflector, as in Fig.4.18b, in which case photons are injected from the side, again via an anti-reflection coating.

An avalanche photo-diode structure of the "reach through" type [4.63] is shown in Fig.4.19. Observe that, in principle, the structure is similar to the one in Fig.4.18a, except for the additional high-field p-region that is

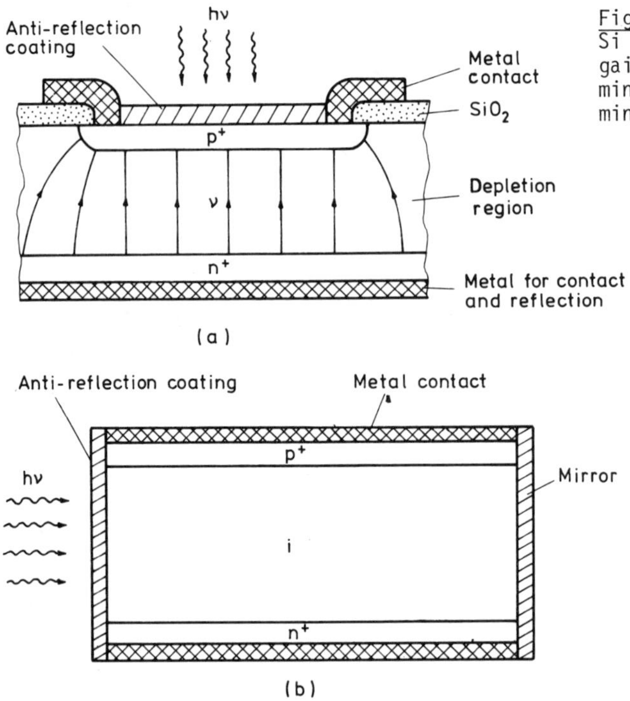

Fig. 4.18a,b. Structure of Si photo-diodes without gain [4.63]. (a) Front-illuminated PIN, (b) side-illuminated PIN

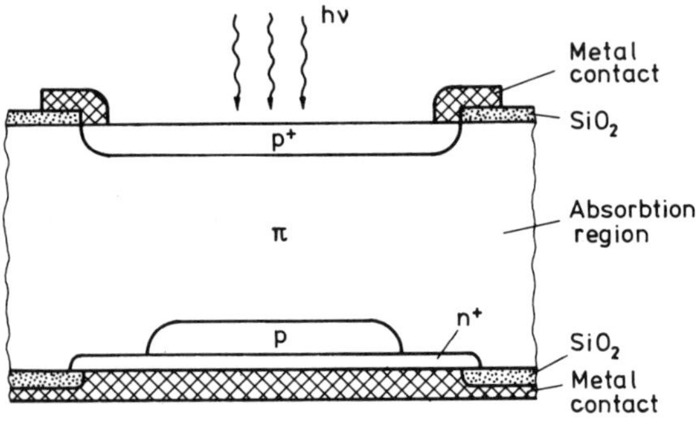

Fig. 4.19. Structure of a typical Si avalanche photo-diode of the reach-through variety [4.63]

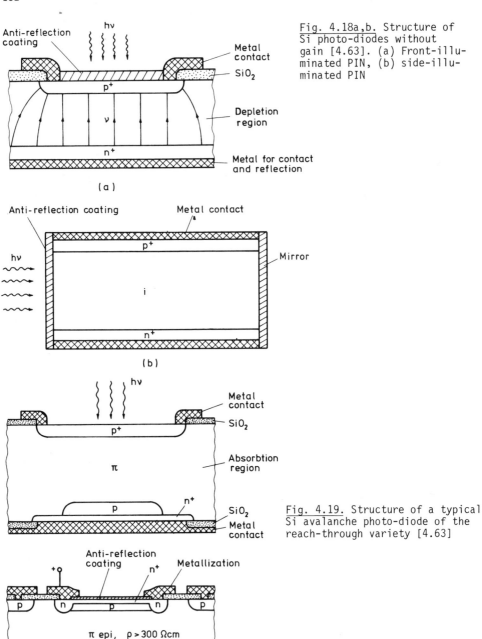

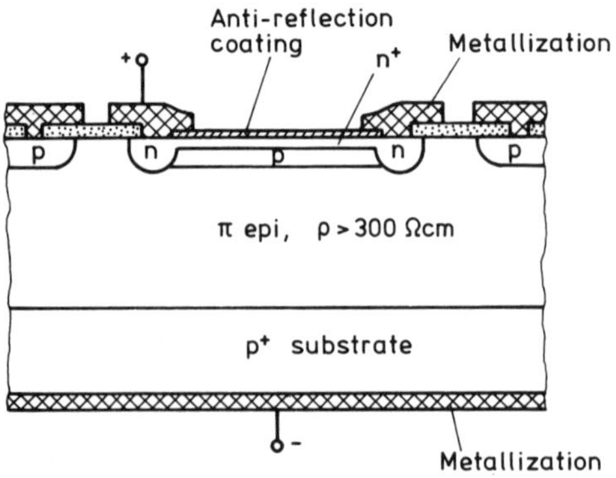

Fig. 4.20. Structure of an epitaxial Si reach-through avalanche photo-diode [4.62]

used for avalanche multiplication. Under low reverse bias, most of the voltage is dropped across the p-n^+ junction. As the bias is increased, the depletion region at this junction broadens and begins to extend deeper into the p region, and eventually "reaches through" to the almost intrinsic π region.

Another structure that is more amenable to fabrication on large silicon wafers [4.62] is shown in Fig.4.20. Here, the layers have been grown on a p^+ silicon substrate and are inverted in comparison to those in Fig.4.19. The incident light now enters through the n^+ contacting layer and, once again, is mostly absorbed in the almost intrinsic π layer. The lightly doped n region forms a "guard" ring that eliminates edge breakdown around the shallow p-n^+ junction. Similarly, the p-type "channel stop" surrounding the device, prevents surface inversion and maintains a low leakage current [4.62]. For more detailed discussions, we refer the reader to the cited works, and to the review articles in [4.63-65].

Let us next consider the external behaviour of PIN and APD photodetectors. Ignoring noise for now, the conditions under which the diodes have to operate are determined by the electronic circuit. A typical circuit configuration (simplified) is shown in Fig.4.21. The photo-diode D is reverse biased to a voltage V_B via the resistor R_B. The effective shunt resistance and capacitance of the diode are marked R_D and C_D, respectively. The output of the diode is coupled via C (for bias isolation) to an amplifier whose input resistance and capacitance are R_A and C_A, respectively. A feedback "block" has been included, to emphasize the fact that the multiplication factor of APD's (Sect.2.3) is a temperature dependent parameter, and must usually be stabilized by feedback. In any case, we can now make a number of observations about the external behaviour of the diode:

1) The speed of response is partly determined by the diode and partly by the circuit. The limitations imposed by the diode are due to the transit time of the carriers, as well as the capacitance of the depletion region. The transit time can be minimized by ensuring that electrons and holes attain their terminal velocities ($\sim 8\times10^4$ ms^{-1} for electrons and 4×10^4 ms^{-1} for holes), and by selecting as narrow a depletion region as possible. Unfortunately, the latter implies an increase in the depletion region capacitance, and a reduction in quantum efficiency. In APDs, we also have an additional delay due to the finite time required for avalanche build-up. However, in practice, the avalanche build-up time is of the order of 0.5 ps [4.66] and is usually negligible in comparison to the delays caused by transit-time and capacitance effects.

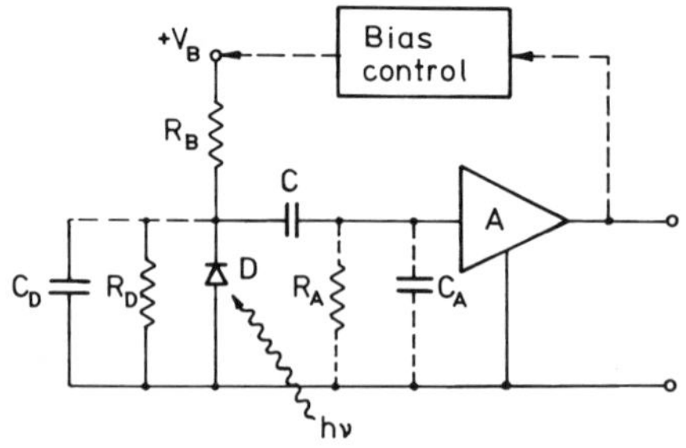

Fig. 4.21. A simplified circuit for opto-electronic conversion using PIN or avalanche photo-diodes

2) For very low light levels, the current flow through the diode will also be low, and there will be no significant change in the voltage across the diode. However, with increasing optical powers, the diode voltage will eventually decrease. In PIN diodes, small changes in the applied voltage are of no importance, but in APDs such changes can have a significant effect on multiplication, as apparent from (2.48). Thus, the value of R_B should be sufficiently low. Unfortunately, even after such a precaution, space charge effects may still lower the junction voltage, while heating effects due to electrical dissipation, as well as optical absorption, may produce temperature dependent changes. In practice, APDs show good linearity up to an optical power level of a few microwatts, but gradually begin to saturate above this level. Further details of saturation effects can be found in [4.67].

3) It should also be noted that the higher the multiplication, the longer the avalanche process persists. Thus, if the optical radiation contains some rapid modulations, they can only be observed at lower multiplication levels. The bandwidth calculations, reported by EMMONS [4.68], indicate that the frequency behaviour of multiplication is characterized by a constant gain-bandwidth product, so that the frequency-dependent multiplication factor $M(\omega)$ can be described by

$$M(\omega) \approx \frac{M(0)}{\sqrt{1+\omega^2\tau^2M^2(0)}} \quad , \qquad (4.9)$$

where ω is the angular frequency of modulation, τ is the effective time of transit through the avalanche region, and $M(0)$ is the zero frequency value of $M(\omega)$.

Let us now turn to the noise behaviour of photodiodes, which, as will be seen in Sect.6.1, is a feature of fundamental importance from the viewpoint of systems design. Basically, noise is a consequence of the randomness of the processes that are involved in the generation of the final output current from the detector. Thus, the probabilistic process of photon absorption is one source of noise (shot noise), while avalanche multiplication is another. Similarly, the random thermal agitation of charge carriers produces "thermal" noise, while random recombinations at traps on the surface of the semiconductor contribute towards "flicker" noise. For the high quality diodes used in optical fiber communications, thermal and flicker noise contributions are usually negligible. However, from the viewpoint of the complete receiver, the thermal noise of the diode load (Fig.4.21) as well as all noise contributions from the amplifier are, in general, significant, and must be accounted for (Sect.6.1).

Let us assume that the quantity we are interested in is the output current of the detector. The output consists of the true signal i_o, and a randomly fluctuating part n, namely,

$$i = i_o + n \quad . \tag{4.10}$$

Unfortunately, there is no way to precisely determine the required parameter i_o. All we can hope is to determine the mean value of i, and associate a variance or standard deviation to it. If the mean of the fluctuating part is zero, the variance of i becomes

$$\mathrm{var}\{i\} = \overline{(i - i_o)^2} = \sigma_n^2 \quad , \tag{4.11}$$

where the bar represents the averaging operation. The relative accuracy of the observation can then be conveniently described in terms of the signal-to-noise ratio, as follows

$$\kappa = \frac{(\bar{i})^2}{\mathrm{var}\{i\}} \approx \frac{i_o^2}{\sigma_n^2} \quad . \tag{4.12}$$

Let us first consider the PIN diode and once again observe that optical radiation causes the flow of a current i(t) that consists of numerous tiny "events" of charge pair generation. The current can thus be written as a summation of these events:

$$i(t) = e \sum_{\ell=1}^{K} h(t - t_\ell) \quad , \tag{4.13}$$

where e is the electronic charge and h(t) represents the response of the diode to a current impulse. In the above, we have assumed that h(t) is constant (not true in general), and that

$$\int_0^\infty h(t)dt = 1 \quad . \tag{4.14}$$

At this point, we are interested in the spectral density of i(t), because this will allow us to evaluate the noise power within any given band [4.69]. Assuming a finite observation duration T, such that $0 \leq t \leq T$, we can evaluate the Fourier transform of i(t) to be

$$\begin{aligned} I_T(j\omega) &= \int_0^T i(t)e^{-j\omega t}dt \\ &= eH(j\omega)\sum_{\ell=1}^{K} e^{-j\omega t_\ell} \quad , \end{aligned} \tag{4.15}$$

where $H(j\omega)$ is the Fourier transform of h(t). We have also assumed that the transit time of the charge $eh(t-t_\ell)$ is short in comparison to T.

By definition, the spectral density of i(t) is given by

$$S_i(\omega) = \lim_{T\to\infty}\left[\frac{\overline{|I_T(j\omega)|^2}}{T}\right] \quad , \tag{4.16}$$

so that we can evaluate the term in the brackets by applying (4.15), as follows

$$\frac{\overline{|I_T(j\omega)|^2}}{T} = \frac{e^2|H(j\omega)|^2}{T} \cdot \overline{\sum_{\ell=1}^{K}\sum_{\ell'=1}^{K} e^{-j\omega t_\ell} \cdot e^{j\omega t_{\ell'}}} \quad . \tag{4.17}$$

The average in (4.17) is most conveniently accomplished in two steps: first, over all possible times $t_\ell \in [0,T]$, and second, over all numbers of events K. Assuming that the events $\{t_\ell\}$ are all statistically independent, and equally probable, we obtain

$$\frac{\overline{|I_T(j\omega)|^2}}{T} = \frac{e^2|H(j\omega)|^2}{T}\left[\bar{K} + \overline{\sum\sum_{\ell\neq\ell'}} \; \overline{e^{-j\omega t_\ell}} \cdot \overline{e^{j\omega t_{\ell'}}}\right] \quad . \tag{4.18}$$

Replacing $\bar{K}$ by IT/e, and letting $T\to\infty$, yields

$$S_i(\omega) = \lim_{T\to\infty}\left[\frac{e^2|H(j\omega)|^2}{T}\left(\frac{IT}{e} + \frac{I^2T^2}{e^2}\,\mathrm{sinc}^2 fT\right)\right] = [eI + I^2\delta(f)]|H(j\omega)|^2 \quad . \tag{4.19}$$

Reference to Fig.4.21 shows us that the DC term $I^2\delta(f)$ will be removed by the coupling capacitor, so that the interesting part of spectral density is eI. Moreover, at the frequencies of interest $H(j\omega) = H(0) = 1$ because of (4.14), so that the AC portion of the spectral density simply becomes

$$S_i(\omega)_{AC} = eI \quad . \tag{4.20}$$

Referring once again to Fig.4.21, we conclude that the bandwidth will be limited by the total shunt capacitance across the diode. Thus, by defining an effective noise bandwidth B_N, and integrating (4.20) from $-B_N$ to $+B_N$, we obtain the well-known Schottky formula

$$P_{N,i} = 2eIB_N \quad . \tag{4.21}$$

The most important conclusion to be drawn from this formula is that the noise power is *signal-dependent*, since the current I is directly related to the optical power. Hence, if the diode is used for detecting optical pulses, the noise power at the peaks will be more than the noise power in the valleys. The formula also reflects the quantized nature of charge and electro-magnetic radiation. To see this, let us write the signal-to-shot-noise ratio in terms of charge, with the aid of the following expressions

$$Q = \int_0^T i(t)dt \quad , \tag{4.22}$$

$$\therefore \bar{Q} \approx IT \quad . \tag{4.23}$$

Also,

$$\mathrm{var}\{Q\} = \int_0^T \int_0^T [i(t) - I][i(t') - I]dt\, dt'$$

$$= \int_0^\infty S_i(\omega)_{AC} \frac{|1 - e^{-j\omega T}|}{\omega^2} \frac{d\omega}{2\pi} = eIT \quad . \tag{4.24}$$

Using (4.23) and (4.24), we obtain

$$\kappa = \frac{\bar{Q}^2}{\mathrm{var}\{Q\}} = \frac{IT}{e} \quad . \tag{4.25}$$

In other words, the signal-to-noise ratio is a measure of the number of electrons that are contained in the charge $\bar{Q} = IT$. Moreover, we know from Sect.2.3 that I and the optical power P are related by

$$I = \frac{\eta e P}{h\nu} = \frac{\eta e \lambda P}{hc} \quad , \tag{4.26}$$

so that substituion into (4.25) gives us

$$\kappa = \eta \frac{E}{h\nu} \quad . \tag{4.27}$$

Setting $\eta = 1$, the signal-to-noise ratio now becomes a measure of the number of quanta that are contained in the energy E. This is the reason why shot noise is sometimes referred to as quantum noise.

We should now consider the more general case of an APD, and include the effect of the randomness of multiplication. If m is the statistically varying multiplication factor, the mean value of the multiplied output current becomes

$$\overline{i_m} = I_m = I \cdot \overline{m} = IM \quad . \tag{4.28}$$

The spectral density, and hence the variance, can be obtained by repeating our previous analysis, with $h(t-t_\ell)$ in (4.13) replaced by $m(t_\ell)h(t-t_\ell)$. Carrying the analysis through shows that the noise power is now given by

$$P'_{n,i} = 2eI\overline{m^2}B_N \quad , \tag{4.29}$$

where $\overline{m^2}$ is the mean-square value of the multiplication factor. This expression can also be written in an equivalent form

$$P'_{n,i} = 2eIM^2F(M)B_N \quad , \tag{4.30}$$

where F(M) is an excess noise factor that accounts for all deviations from the ideal (noise-less) multiplier, in which $\overline{m^2} = \overline{m}^2$. The value of F(M) is affected by a number of factors, including the details of the diode structure and the degree to which only one carrier type (electron or hole) is multiplied. McINTYRE [4.70] has considered the complex problem of multiplication statistics in uniform diodes, and gives the following simplified expression for F(M)

$$F = k'M + (2 - 1/M)(1 - k') \quad , \tag{4.31}$$

where k' is the ratio of the ionization rates of holes and electrons (respectively) for electron injection, and vice versa for hole injection. Clearly, we would like as small a value of F as possible, so that k' should also be minimized. Another consideration is related to the assumption of a uniform electric field, which rarely holds true in practice. However, the expression in (4.31) remains useful if k' is replaced by an effective (weighted) value [4.70]. For typical silicon APD's, the effective value of k' may be about 0.02, implying an excess noise factor of 3.95 at M = 100. In comparison, the best achievable value is given by F = 2, for k' = 0 and large M.

Finally, to appreciate the benefits of multiplication, we should also consider the thermal noise generated by the resistance which loads the APD. The energy of the noise can be deduced from the thermo-dynamical equipartion of energy kT/2, which is associated with each degree of freedom (k is Boltzmann's constant, and T is the absolute temperature).

Suppose that a resistor R is connected in parallel with an ideal capacitor C, as shown in Fig.4.22. The thermal noise of the resistor is represented by an equivalent source that has a spectral density $S_i(\omega)$, and a mean-square current $\overline{i^2}$. This generator drives the RC network and produces a voltage V across the capacitor. The resultant stored energy must also equal kT/2, so that we have

$$\frac{1}{2}CV^2 = \frac{1}{2}kT \quad . \tag{4.32}$$

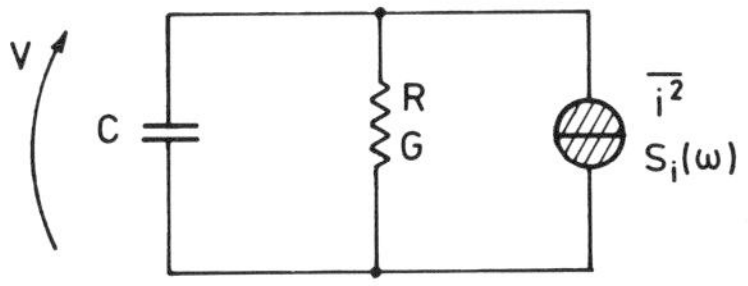

Fig. 4.22. An RC circuit for the derivation of thermal noise

The voltage V can be easily evaluated by using the transfer function of the circuit, and by taking $S_i(\omega)$ to be a constant. This latter assumption is justified because thermal movements are very rapid and produce a spectrum that is approximately constant at the frequencies that are of interest. Using this approach, we have

$$\begin{aligned} \frac{1}{2}CV^2 &= \frac{C}{2}\int_{-\infty}^{\infty} \frac{S_i(\omega)}{|G + j\omega C|^2} \cdot \frac{d\omega}{2\pi} \\ &= \frac{S_i(\omega)}{4\pi C}\int_{-\infty}^{\infty} \frac{d\omega}{\omega^2 + G^2/C^2} \quad , \end{aligned} \tag{4.33}$$

where the total power is computed by integrating the power density of V over all frequencies (angular frequencies/2π). Performing the integration and using (4.32), we obtain

$$S_i(\omega) = 2GkT \quad . \tag{4.34}$$

This corresponds to a mean-square value of

$$\overline{i^2} = 4GkTB_N \quad , \tag{4.35}$$

obtained, as before, by using the equivalent noise bandwidth B_N. Using (4.28), (4.30), and (4.35), the overall signal-to-noise ratio becomes

$$\kappa = \frac{M^2I^2}{(2eM^2FI + 4GkT)B_N + \overline{i_N^2}} \quad , \tag{4.36}$$

where $\overline{i_N^2}$ is the equivalent noise contribution of the amplifier or observation instrument. Note that background and dark currents can also be included in the current I.

We now see that multiplication may allow us to reduce the relative significance of both thermal and amplifier noises, particularly for large signals and/or multiplication factors. As will be seen in Sect.6.1, a large multiplication factor presupposes a low excess noise factor. Finally, it should be noted that (4.36) can also be used for PIN diodes if M, and hence F, are equated to unity.

4.5 Coupling

In the previous chapter we have discussed the various elements that are required for the construction of an optical fiber system. In preparation for Chapter 6, in which we consider applications, we should now take a look at some of the means available for connecting the various elements together.

First, we should note that the "factory length" of optical cables is generally restricted, and that for communication links, several cable sections normally have to be coupled together to form the complete channel. Thus, our first problem is the coupling of nearly identical fibers. The second problem is to couple modulated optical power to the input of the "jointed" channel, while the third is to couple the fiber's output power to a detector.

Starting with the last problem, we observe that, in practice, the area of detectors is usually larger than the area of the fiber cores, so that it is conceptually a straight-forward matter to couple power from a fiber to a detector. Coupling loss arises mostly from Fresnel reflections, and can be minimized by refractive index matching, or by the use of an anti-reflection layer on the detector. Except for these observations, we intend to ignore this aspect of the coupling problem.

Fiber-to-fiber joints can be classified into two categories: 1) permanent or semi-permanent joints achieved by bonding or thermal fusion, and 2) detachable joints using "de-mountable" connectors. In commercial communications systems, most of the joints are usually of the first type, because of the higher reliability and stability that they offer under field conditions. However, from the viewpoint of maintenance and service, it is more reasonable to use de-mountable connectors for coupling the channel to the terminal equipment. A popular approach is to permanently connect a short fiber "pig-tail" to the optical source, and to use a connector between the pig-tail and the actual channel. The same solution is also useful at the detector end, while in low bit rate systems the detector can also be included in the connector itself.

The demands on "panel" connectors are not as severe as on those intended for field use, because of the relatively "friendly" environment within which terminal equipment is normally located. On the other hand, field connectors are required to operate over a wide range of temperatures and humidities. This is one reason why fiber-to-fiber *fusion* splices are very attractive, because not only are the materials thermally matched, but interfaces into which water can creep are completely avoided.

One of the earliest demonstrations of a thermally fused splice was based on welding by a hot wire [4.71]. However, modern silica fibers, with their relatively high fusion temperatures (1600° C), are more conveniently welded by using a confined electrical arc [4.72-74]. Fusion machines, based on this technique, are commercially available from a number of manufacturers, and offer average splice losses that are better than 0.1 dB. Such low losses presuppose that small defects in the fiber ends are removed by "heat polishing", i.e. by applying the arc to the fiber ends while they are still slightly apart ($\sim 1\ \mu m$) [4.75]. In addition, an important mechanism that aids in the achievement of low splice losses is the existence of surface tension at the moment of fusion. This surface tension tends to correct for small mechanical misalignments of the fibers [4.76], and becomes particularly useful for the fusion of single-mode fibers [4.77]. Splice losses of about 0.1 dB can also be achieved in this case. The main drawback of the fusion method is a possible reduction in the fiber strength in the vicinity of the splice [4.72]. However, it is normally straight-forward enough to protect these weak regions.

Before leaving the question of splices, we would like to draw the reader's attention to a number of other splicing methods that are worth considering. Most of these are based on the mechanical end-to-end alignment of fibers using a diversity of means such as embossed or etched V-grooves [4.78], capillary tubes [4.79], "loose" metal tubes [4.80], etc. A linear array of V-grooves is also convenient for the splicing of fibers in ribbon format [4.81], particularly when the grooves are obtained by the accurate preferential etching of silicon substrates [4.82]. For further details, we refer the reader to the review in [4.83].

The other type of joint that was mentioned earlier was the de-mountable connector. Every major manufacturer of conventional connectors has produced *at least* one design, while many "in-house" designs have been produced by the large number of telecommunications companies involved in optical fiber communications. We therefore feel justified in ignoring the technological world of connectors, and will refer the reader to [4.5] and [4.84-89], as well as to the references therein, for an overview of some published designs.

Instead, we will address ourselves to the question of source-to-fiber coupling which, besides being instructive, also tends to be a serious source of loss in fiber optical systems. If the source would have a transverse field distribution that exactly matches the transverse field distribution of the fiber, the coupling problem would be more or less trivial and the fiber could be simply "butt" jointed to the source. However, in general the field distributions tend to be different both in shape and in size. In such cases, we may attempt to transform the source distribution into the form of the fiber distribution, by using various optical elements such as lenses.

Whether such a transformation is used or not, what finally interests us is the efficiency with which power can be launched into the fiber. For a single-mode fiber, we must go through a field analysis to evaluate the launching efficiency [4.90-92]. Of course, we could do the same for multimode fibers and evaluate the coupling efficiency of each guided mode, but that would be a tedious task indeed. Fortunately, typical multimode fibers support several hundred modes, so that geometrical optics can be used without any great decrease in accuracy. Moreover, because single-mode fiber systems are not treated in this book, we will restrict ourselves to the geometrical optics - multimode fiber case.

However, before doing so, we should first establish some criteria for deciding whether or not the use of lenses will improve the coupling efficiency. For this purpose, we will invoke Liouville's theorem [4.93,94], which for the fiber-source problem is most conveniently written in the following form

$$dV = n^2 dA d\Omega = \text{constant} \quad , \tag{4.37}$$

where dV is the differential volume in phase space (coordinate-propagation vector), dA is the emission surface, $d\Omega$ is the solid angle of emission, and n is the refractive index of the surrounding medium. The implication of (4.37) is the following: any volume in phase space, which contains a certain number of representation points describing the position and angle of a given number of propagation vectors, must always remain constant. In other words, if the phase volume is established by the radiation source, it is impossible to decrease it by optical processing without excluding some representation points (i.e. power). Thus, the maximum achievable coupling efficiency is given by a ratio of the fiber's input phase volume to that of the source. If the phase volume of the source is larger than that of the fiber, the efficiency will be less than unity, and can only be improved by changing the source or the fiber.

Let us suppose that an ideal Lambertian source of surface area A_s (Fig. 4.23) is to be coupled to a step-index fiber of core area A_f, and numerical aperture NA. The solid angle into which the source radiates is, of course, 2π steradians, while the cone within which the fiber accepts radiation is very easily shown to have a solid angle of $2\pi(NA)^2$. Hence, application of (4.37) gives us the maximum achievable efficiency

$$\eta = \frac{A_f}{A_s} \cdot (NA)^2 \quad . \tag{4.38}$$

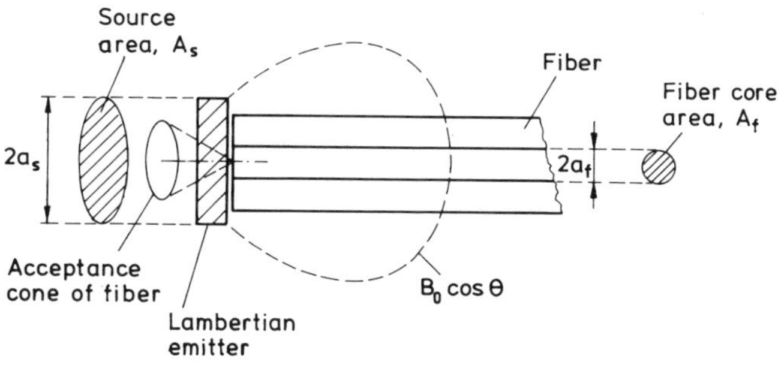

Fig. 4.23. Schematic representation of power launch from a Lambertian source into a step-index fiber, for the illustration of Liouville's theorem

Since LEDs are very nearly Lambertian sources, (4.38) also represents the upper limit on LED-to-fiber oupling efficiency. In order to compare this to some realistic values, let us assume that a step-index fiber is directly "butted" against an LED. If we assume a circular emission area that has a radiance (Watts per square meter steradian) $B(\theta) = B_0 \cos\theta$, then the power collected by a fiber of arbitrary profile, can be written in the form [4.95]

$$P_f = 4\pi^2 B_0 \int_0^{r_c} \int_0^{\theta_0(\rho)} \rho \cos\theta \sin\theta \, d\theta \, d\rho \quad . \tag{4.39}$$

where ρ is the radial position, r_c is the radius of the area common to both the source and the fiber, and $\theta_0(\rho)$ is the maximum acceptance angle of the fiber at ρ. The integral is easily evaluated over θ with the result

$$P_f = 2\pi^2 B_0 \int_0^{r_c} [n^2(\rho) - n^2(a)]\rho d\rho \quad . \tag{4.40}$$

where $n(\rho)$ and $n(a)$ are the refractive indices of the core and cladding regions, respectively. The total power emitted by the source is simply

$$P_s = \pi^2 r_s^2 B_0 \quad , \tag{4.41}$$

so that the coupling efficiency becomes:

$$\eta = \frac{2}{r_s^2} \int_0^{r_c} [n^2(\rho) - n^2(a)]\rho d\rho \quad . \tag{4.42}$$

Here r_s is the radius of the LED. For a step-index fiber, (4.42) reduces to:

$$\eta_1 = (r_c/r_s)^2 (NA)^2 = \frac{A_c}{A_s} (NA)^2 \quad . \tag{4.43}$$

Let us now indentify two regions of interest: (1) the fiber area $A_f \leqq A_s$, and (2) $A_f > A_s$. In the first case $A_c = A_f$, and we obtain the result:

$$\eta_1' = \frac{A_f}{A_s} (NA)^2 \quad \text{for} \quad A_s \geqq A_f \quad . \tag{4.44}$$

This is seen to be identical to (4.38), and we conclude that, when the LED area is greater than or equal to the core area of the fiber, the butt joint immediately yields the maximum achievable efficiency, and the situation cannot be improved by the use of lenses.

In the second case, when $A_s < A_f$, the common area A_c becomes equal to A_s, and the coupling efficiency becomes:

$$\eta_1'' = (NA)^2 \quad . \tag{4.45}$$

Since $A_s < A_f$, the corresponding maximum achievable efficiency is greater than η_1'', so that we should be able to improve the coupling efficiency. Apparently, we should project a magnified image of the source onto the fiber, such that the image area would be equal to the core area. However, we should bear in mind that coupling efficiency and the actual power launched into the fiber are two different parameters. Ultimately, we would like to maximize the input power, but in practice reduction of the source area may be accompanied by a lower output power from the device. In such cases, the most reasonable compromise may be to simply match the areas of the LED and the fiber.

It is also worth noting that the integral in (4.42) can be easily evaluated for the case of the parabolic index profile, with the result

$$\begin{aligned} \eta_p &= (A_f/A_s)(NA)^2_{max}/2 \qquad \text{for} \quad A_s \geq A_f \quad , \\ &= [1 - (A_s/A_f)/2](NA)^2 \quad \text{for} \quad A_s < A_f \quad , \end{aligned} \tag{4.46}$$

where $(NA)_{max}$ is the numerical aperture at the core centre. Thus, for $A_s \geq A_f$, the coupling efficiency is half that of the step-index case. This is rather understandable since a parabolic fiber supports only half the propagating modes of a geometrically identical step index fiber. On the other hand, for $A_s < A_f$, η_p is more complex, but approaches η_1'' for $A_s << A_f$.

When a small area LED is used in conjunction with image magnification for area matching, in principle the radiance characteristic is narrowed from its original cosine form. However, for the small acceptance angle of weakly guiding fibers, the change in the radiance is normally insignificant, and (4.38) and (4.44) can still be used for moderate magnifications. For large magnifications, changes in the radiation pattern and aberrations in the imaging system must also be taken into account.

Although our foregoing considerations are related to Lambertian sources, they allow us to make the following generally applicable deductions:

1) When both the source area and its numerical aperture (divergence) are larger than the core area and numerical aperture of the fiber, the source-to-fiber coupling efficiency cannot be improved by the use of lenses.

2) When the source area is less than the core area of the fiber, but the numerical aperture is larger than that of the fiber, we can project a magnified image whose area matches that of the core and thereby improve the coupling efficiency. In other words, we can reduce the divergence at the expense of increasing area.

3) When both the area and the numerical aperture of the source are less than those of the fiber, we can obviously obtain 100% coupling efficiency (neglecting reflection losses, etc.) by simply butting the fiber against the source.

4) When the source has a larger area than that of the fiber core, but its numerical aperture is smaller, we can de-magnify the image and match the numerical apertures of the fiber. This is the situation that is mostly encountered in the use of gas or solid-state lasers, particularly in experimental work. Hence, in such cases, the use of a de-magnifying lens is advisable.

5) When a source is coupled to a graded-index fiber, we should bear in mind that the numerical aperture approaches zero near the cladding. In such cases it may be advisable to restrict the effective (imaged or otherwise) source area to somewhat less than that of the core, depending on divergence considerations.

Let us now apply some of these thoughts to the problem of coupling between DH stripe lasers (Sect.4.2) and fibers. First, let us note that the beam profile of typical laser diodes is usually asymmetric (Fig.4.24), and has a near elliptical area. The divergence in the plane perpendicular to the junction is usually larger than in the plane parallel to the junction. Normally, the latter divergence is small (5°-10°) and creates no problem. However, perpendicular to the junction, the divergence is usually significantly larger (10°-30°) than the acceptance angle of typical multimode fibers. On the other hand, the length of the major and minor axes of the near-field ellipse (not the junction) may be in the order of 20 μm and 2 μm, respectively, or at least can be made so by suitable diode design. Bearing in mind that typical multimode fibers have diameters of about 50 μm, we see that high coupling efficiencies could be achieved by selectively magnifying the narrow emission region. For example, the use of two cylindrical lenses (Fig.4.25) allows us to transform the elliptical emission region into a circular image. Alternatively, a single cylindrical lens could be used, followed by a spherical lens for generating an image of the correct size. Yet another alternative that, incidentally, helps to reduce reflection losses, is the use of slabs with

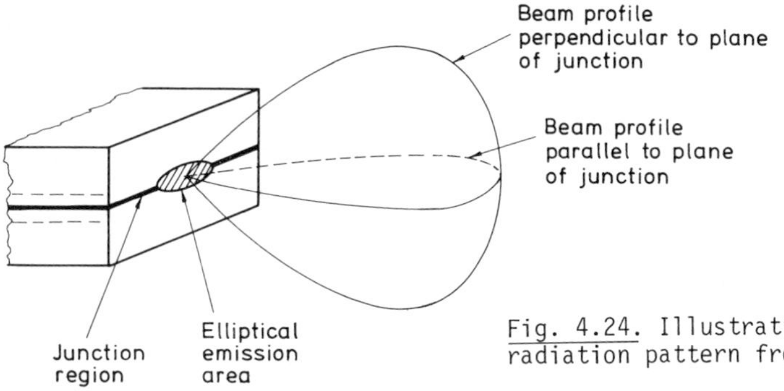

Fig. 4.24. Illustration of an idealized radiation pattern from a DH stripe laser

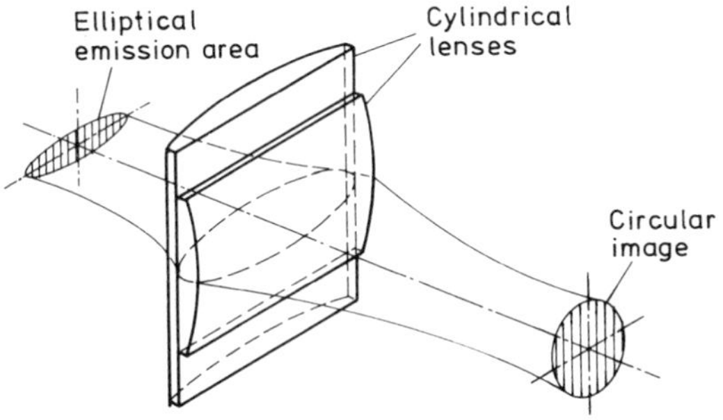

Fig. 4.25. Transformation of an elliptical emission region into a circular image

parabolic refractive index profiles. These slabs perform as lenses (Sect.3.4), but being flat, they can be bonded with index-matching cement. Of course, from the viewpoint of power collection by a multimode fiber, an elliptical-to-circular transformation is not strictly necessary, because all we would like to do is to reduce excessive divergence. However, it helps to soften the crude approximation we intend to make in the following.

We will assume that we effectively have a circular image at the fiber-end face, which launches power into the fiber within a numerical aperture $(NA)_s$ corresponding to a half angle θ_s. We will also assume that the radiance within the core is a constant B_0, and that the image area is greater than or equal to the core area. Then application of (4.37) tells us that the maximum achievable coupling efficiency for a step index fiber is given by

$$\eta_2 = \frac{A_f}{A_s}\frac{(NA)_f^2}{(NA)_s^2} \quad , \qquad (4.47)$$

where the subscripts f and s refer to fiber and the image of the source, respectively. Clearly, if we set $A_f = A_s$ and $(NA)_f = (NA)_s$, a coupling efficiency of 100% can be achieved. On the other hand, the efficiency is only 50%, i.e. 3 dB loss, for a parabolic fiber.

Let us next evaluate the efficiency using geometrical optics, as in the case of the LED. We will assume that all the power available from the source is also available at the image. Then, using (4.39) we obtain

$$P_s = 4\pi^2 B_0 \int_0^{r_s} \int_0^{\theta_s} \rho \sin\theta \, d\theta \, d\rho \quad ,$$

$$= 2\pi^2 B_0 r_s^2 (1 - \cos\theta_s) \quad . \tag{4.48}$$

Similarly, the power accepted by the fiber is given by

$$P_f = 4\pi^2 B_0 \int_0^{a} \int_0^{\theta_0(\rho)} \rho \sin\theta \, d\theta \, d\rho$$

$$= 4\pi^2 B_0 \int_0^{a} \{1 - \sqrt{[1 - n^2(\rho) + n^2(a)]}\}\rho \, d\rho \quad . \tag{4.49}$$

For weakly guiding fibers, this integral can be simplified, and we obtain the following coupling efficiency

$$\eta_2 \simeq \frac{1}{r_s^2(1 - \cos\theta_s)} \int_0^{a} [n^2(\rho) - n^2(a)]\rho \, d\rho \quad . \tag{4.50}$$

In the case of step-index fibers, (4.50) reduces to

$$\eta_2'' = \frac{A_f}{A_s} \frac{[1 - \cos\theta_0(0)]}{[1 - \cos\theta_s]} \quad ,$$

$$\approx \frac{A_f}{A_s} \cdot \frac{(NA)_f^2}{(NA)_s^2} \quad . \tag{4.51}$$

Thus, a coupling efficiency of 100% can be achieved for step-index fibers. Similarly, (4.50) can be evaluated, as earlier, for the parabolic refractive index profile, and in the weakly-guiding approximation, the following result is obtained

$$\eta_2 \approx \frac{1}{2} \frac{A_f}{A_s} \frac{(NA)_f^2}{(NA)_s^2} \quad . \tag{4.52}$$

We see that both for step-index and parabolic fibers, the results given in (4.51) and (4.52) are identical to the ones obtained by the application of Liouville's theorem.

In practice, the approach to the source-fiber coupling problem is to fabricate as small a device as is consistent with the available technology, and to use an optical "transition" to increase the effective numerical aperture presented to the source. Examples of such transitions are the bulb-ended fiber [4.96-98], and the fiber taper [4.99]. The former can be formed by melting the end of the fiber, mostly after removing the cladding by etching. Surface tension then helps to produce a spherical lens at the end of the fiber. Coupling efficiencies up to about 9% have been reported for coupling to LEDs [4.97], and about 50% for coupling to laser diodes [4.96].

The fiber taper can be formed by heating and pulling the fiber [4.99], and is the more interesting device, both because of the inherently higher efficiency that it offers, and the flexibility with which the taper ratio can be changed. Moreover, the taper can also, in principle, be fabricated to match the elliptical near-field profile of semiconductor lasers, to the circular profile of fibers.

The operation of the taper can be easily understood by following the ray trajectory shown in Fig.4.26, from which it can be seen that as the ray progresses along the taper, its angle of incidence gradually increases. In other words, the numerical aperture at the narrow end of the taper must be larger than at the fiber end. The extent of the increase can be estimated either by geometrical optics, or more simply by once again using Liouville's theorem, as discussed below.

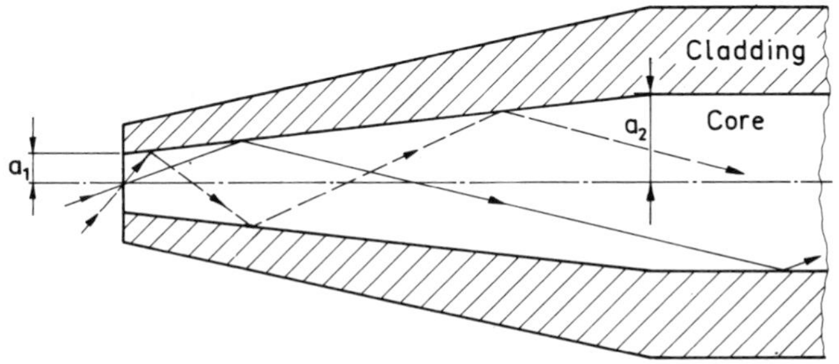

Fig. 4.26. Principle of operation of fiber taper

If we define an acceptance cone at the narrow end of the taper, such that all radiation incident within this cone is accepted by the fiber, it follows that the phase volume of the input cone must equal the phase volume of the fiber. Then, using the form of the solid angle that led us to (4.38), we can write

$$\pi a_1^2 \cdot 2\pi(NA)_1^2 = \pi a_2^2 \cdot 2\pi(NA)_f^2 \quad ,$$

or

$$(NA)_1 = \frac{a_2}{a_1}(NA)_f = R \cdot (NA)_f \quad . \tag{4.53}$$

Here, a_1 and a_2 are the radii of the narrow and broad ends of the taper, respectively, while $(NA)_1$ is the effective numerical aperture at the narrow end. Assuming that the narrow end of the taper is butt jointed to a LED of the same dimensions, the efficiency of coupling becomes

$$\eta'_{taper} = R^2(NA)_f^2 \quad . \tag{4.54}$$

In other words, the coupling efficiency is increased by a factor of R^2. Similarly, for a laser diode with the same emission area as that of the narrow taper end, (4.47) tells us that the maximum achievable efficiency is given by

$$\eta''_{taper} = R^2 \frac{(NA)_f^2}{(NA)_s^2} \quad . \tag{4.55}$$

Of course, the assumption is that $(NA)_s > (NA)_f$, so that we see that a coupling efficiency of 100% could be achieved for $R(NA)_f = (NA)_s$, corresponding to an improvement factor of R^2. However, in practice, only the divergence perpendicular to the plane of the diode junction is likely to exceed the acceptance of the fiber. In this case, we could reduce the problem to two dimensions, and assume that we have a "line" source of the same size as the diameter of the narrow end of the taper. The phase "volume" of interest is now simply the product of the line width and the numerical aperture (for small NA). The corresponding coupling efficiency becomes

$$\eta'''_{taper} = R \cdot \frac{(NA)_f}{(NA)_{max}} \quad , \tag{4.56}$$

where $(NA)_{max}$ is the numerical aperture perpendicular to the junction. Hence, the improvement factor in the case of a typical DH stripe laser can be expected to be R, instead of R^2. This relationship is in reasonable agreement with the experimental results of OZEKI and KAWASAKI [4.99]. It should also be noted that, although the above improvement factors have been calculated upon the assumption of equal areas or line widths, obviously the same improvement factors are obtained for unequal areas or line widths.

To conclude this section, we will briefly comment on the behavior of the excited modal fields, as they propagate along the fiber. First, we should note that because of geometrical and optical inhomogeneities in the fiber, a certain amount of mode conversion always exists. As a result, irrespective of the modal power distribution at the input of the fiber, a steady-state distribution will be attained after a sufficiently long distance. Hence, if all modes would be equally excited at the input, power from the higher-order mode groups would be gradually pumped into radiating and leaky modes. From the viewpoint of the complete optical fiber link, it would then seem that the source-to-fiber coupling efficiency is lower than that measured on the basis of a short length of fiber. Thus, we should try to excite the lower order modes at the input of the fiber. Unfortunately, this has another detrimental influence that will be discussed in Chap.5.

A similar behavior can also be expected at every fiber-to-fiber joint at which there is a mismatch in the refractive index profiles, or a mechanical mis-alignment (angular, transversal, or longitudinal). It is almost obvious that differences in refractive index will produce angular changes that correspond to changes in the modal order. As a result, propagating modes may be converted into leaky or radiating modes. Similarly, mechanical mis-alignments will also produce changes in the modal order. A loss is, of course, inherent if power is coupled into leaky and radiating modes, but, as before, an additional loss is also incurred because of deviations from the steady-state modal power distribution of the following fiber.

5. Fiber Measurements

In the use of fibers for communication purposes, a factor of prime interest is the extent to which a transmitted signal changes during propagation. Bearing in mind that such changes are introduced by power loss and material and modal dispersions, it becomes clear that ultimately we must have accurate knowledge of the attenuation within the system, as well as the system's impulse response (or its Fourier transform, the frequency response). At first sight, we are left with a seemingly straightforward measurement problem. Unfortunately, the situation is complicated by a number of variables which ensure that neither the measured attenuation of a given fiber nor its measured impulse response are universally and uniquely defined quantities. We will therefore devote some space (Sect.5.1) to a short discussion of these problems, before considering measurement techniques.

5.1 General Considerations

Some of the more important factors that affect the attenuation and impulse response of a fiber are summarized in Table 5.1, in which the total attenuation has been split into its components, and both single and jointed fibers have been included for comparison.

The first column shows that the source-to-fiber coupling loss is affected by the source radiation pattern and by the fiber refractive index profile, as well as by the source and fiber geometries. This is easily understood since the fiber index profile affects its numerical aperture, which ensures that only radiation incident within certain angles can propagate in the fiber [5.1]. On the other hand, even when the angular requirements are fulfilled, loss inevitably occurs if the area of the source is larger than the fiber core area. On the whole, however, source-to-fiber coupling attenuation tends to be relatively constant except for the possible influence of source aging, which can lead to changes in the source radiation pattern. This loss

Table 5.1. Indicates the effect of certain factors (1st column on left) on the parameters of interest in an optical fiber system. Two asterisks (**) indicate a major influence, while a single asterisk (*) shows a second-order effect, of little practical importance

	Attenuation				Dispersion	
Influencing Factors	Source fiber coupling loss	Single fiber loss	Jointed fiber loss	Fiber-detector coupling loss	Single fiber	Jointed fiber
(1) Source Radiation Pattern	**	**	**		**	**
(2) Source Geometry	**					
(3) Source Spectral Width					**	**
(4) Source Wavelength		**	**		**	**
(5) Fiber Geometry	**			*	**	**
(6) Fiber Index Profile	**			*	**	**
(7) Launched Mode Distribution		**	**		**	**
(8) Splices or Connectors			**			**
(9) Detector Geometry				*		

can be determined by the straightforward measurement of the ratio of total emitted power from the source to the power carried by propagating modes, for example, in the steady state.

The second and third columns in Table 5.1 relate to the attenuation in single and jointed fibers, respectively. In both cases, the attenuation is wavelength dependent, and also varies somewhat with launching conditions. In jointed fibers, the splice (or connector) obviously introduces additional loss. The measurement of fiber and splice attenuations will be discussed in Sect.5.2.

Before turning our attention to the columns relating to dispersion in Table 5.1, we note that a column on fiber-to-detector coupling loss has been included for the sake of completeness. However, this attenuation is not normally important, because in practice the surface area of detectors is often substantially larger than the area of the fiber core so that power loss is mostly due to Fresnel reflections. Such loss is usually negligible in comparison to the other losses in the system.

Let us now consider the dispersion columns in Table 5.1. The first striking feature seems to be that just about everything has some influence on dispersion. However, the geometry and index profile are factors that are controlled during production and they do not vary during measurement. Similarly, source wavelength and spectral width are kept constant within a given measurement, and these may also be neglected in our present discussion. This leaves us with the influence of the launched mode distribution and, in jointed fibers, with the splice itself. In fact, the influence of the splice is to disturb (by filtering, scrambling or converting) the mode distribution entering the splice, such that the mode distribution at the input of the next fiber is different from the expected one. Even with a perfect splice, differences between the refractive index profiles will also cause mode conversion [5.2]. In effect, then, we can reduce the problem to (I) the determination of the mode distribution at the input of a fiber (with or without a splice), and (II) its influence on fiber dispersion. Given this situation, how should the fiber impulse (or frequency) response be specified? One approach is to always measure it under equilibrium mode excitation conditions. However, this does not necessarily avoid the splice problem, and does not always represent optimal use of the fiber. An alternate approach is to specify selective excitation conditions. In this case, mode equilibrium is reached after a certain distance that depends on the mode mixing properties of the fiber. If a splice is inserted at some point before modal equilibrium [5.3], the situation would require a precise knowledge of the modal behavior of the splice as well as the fibers. It is clear that all these aspects affect the accuracy with which the response of a fiber link can be determined, so that the systems designer must allow for appropriate margins.

Within the context of these difficulties, our purpose in this chapter is to present some basic principles for the determination of attenuation and dispersion, which, as we have discussed, are the actual parameters of interest to the systems designer. However, because the refractive index profile is a major factor in the theoretical assessment of fiber dispersion, we have also included a section on profile determination. On the other hand, we completely ignore the measurement of source characteristics and fiber dimensions, because the source radiation pattern can be determined by using the equipment discussed in Sect.5.5, the spectral properties can be found by using the apparatus described in Sect.5.2, while various source and fiber geometries can be easily determined by using well known microsopic techniques. We recognize the importance of on-line fiber diameter measurements for production control, but consider such measurements to be outside the scope of this book.

Another aspect that will not be discussed in the forthcoming sections, but which is a pre-requisite of all fiber measurements, is a high quality fiber end. The lack of such a fiber end can give rise to scattering that can lead to unexpected mode excitation, and thus to poorer coupling efficiency. The most familiar, perhaps most straightforward, technique for preparing a fiber end face has been described by GLOGE et al. [5.4], and is based on fracturing a curved fiber that is under strain. Figure 5.1 shows a cutting device based on this technique. Simpler machines based on the same principle have also been described [5.5]. An alternative, which is often particularly useful in the laboratory, is to place the fiber end in an index-matching cell (with a glass window). In this way, poorer ends can also be used, and even the scattering effects of large discontinuities can be conveniently matched out.

Fig. 5.1. A cutting device for fibers (Reproduction, publication with permission of Forschungsinstitut der Deutschen Bundespost)

Finally, we would like to remind the reader of the framework within which fiber measurements are performed. First, due to the necessarily small dimensions involved, an abundance of micrometer stages for x, y , z, rotation, and tilt motions are normally necessary for alignment and adjustment. The required optical "circuits" are then formed, as usual, on optical rails and tables. Secondly, most measurements are performed in the infra-red (IR) regime, which complicates the problem of viewing and alignment. The usual answer is to use a portable IR viewer and/or an IR sensitive TV camera and monitor. The latter can also double as an instrument for measurements, as will be seen in Sect. 5.5. Often, it is more convenient to carry out initial alignments using a He-Ne laser instead of the IR source.

5.2 Spectral Attenuation

The most popular method for spectral attenuation measurement involves the observation of the output optical power at the end of a long fiber and its comparison to the power output from a short reference length of the same fiber. The measurement is then repeated at various wavelengths using a monochromator. If we assume that the output from the long length of fiber is $P_1(\lambda)$, and that a higher power $P_2(\lambda)$ is observed after cutting off a length of fiber L, the attenuation can be defined in the usual way as

$$A(\lambda) = \frac{10}{L} \lg \frac{P_2(\lambda)}{P_1(\lambda)} \text{ [dB/km]} \quad , \tag{5.1}$$

where the length L is assumed to be in kilometers. From (5.1) we see that four parameters affect the measurement accuracy: the two measured powers, the wavelength in question, and the fiber length. The first two are affected both by the instrumentation electronics (which should be stable, drift-free, and have adequate noise properties), and by the stability of the source. Thus, if the temperature of the source changes or drifts (e.g. due to power supply fluctuations), the total radiant exitance [W cm^{-1}] changes according to the Stefan-Boltzmann law:

$$W = \sigma T^4 \quad , \tag{5.2}$$

where T is the absolute temperature, and σ is the Stefan-Boltzmann constant. It is thus obvious that the source (lamp) power supply must be carefully stabilized. The third factor, wavelength accuracy, is related to monochromator calibration, and will be neglected here. The last factor that affects the accuracy of measurement is the length L. Perhaps the most convenient means for measuring L is to use the optical time domain reflectometer (OTDR), to be discussed later in this section.

Next we must consider the wavelength region of interest and the range of attenuations which can be measured. The latter is primarily affected by the optical power that can be launched into the fiber (at a given wavelength), and can be improved to some extent, by proper choice of instrumentation. On the other hand, the pertinent wavelength range is mostly determined by what the technology has to offer. Figure 5.2 shows the attenuation curve of a low loss fiber [5.6], from which it can be seen that we should at least cover the range 0.8-1.6 μm. Frequently, two different detectors are used for this purpose, even though a single germanium diode may be sufficient. However, germanium diodes have the disadvantage of being relatively noisy, particularly at room temperature, so that their use entails additional complexity in detection instrumentation. Figure 5.3 shows the block diagram of a measurement set-up that minimizes such noise difficulties, and satisfies the wavelength requirements. Light from a suitable lamp is filtered, chopped into a pulse stream, and fed into the monochromator. The output radiation from the

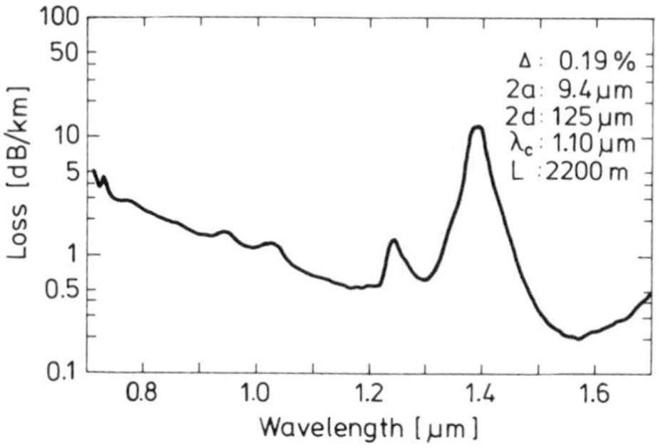

Fig. 5.2. Attenuation of a low-loss fiber [5.6]

Fig. 5.3. Block diagram of the spectral-attenuation measurement set-up

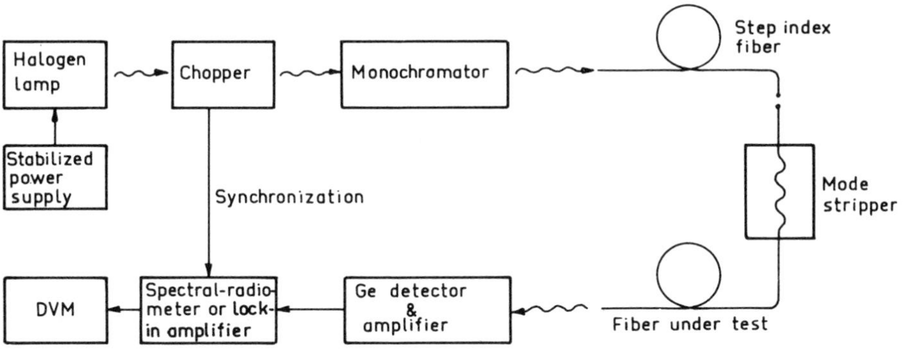

monochromator is coupled into the fiber using a collimator lens and a microscope objective. Alternatively, the light can be first launched into a step-index fiber (with core dimensions similar to or larger than those of the fiber being measured, and a larger numerical aperture), which then acts as a source for the fiber under test. This ensures that the latter fiber is fully excited and that the excitation is relatively insensitive to axial misalignment [5.7] (Fig.5.4). Such excitation leads to repeatable measurement results, but also excites a substantial number of cladding modes, which must be carefully removed, e.g. by immersing the fiber in an index matching liquid.

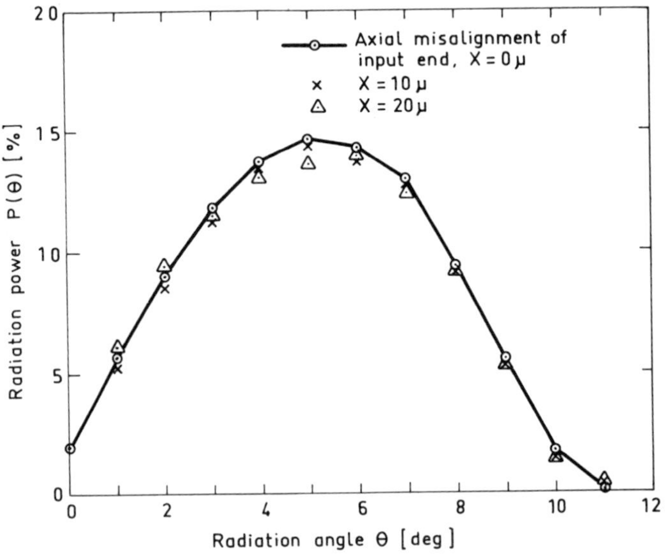

Fig. 5.4. Far-field characteristics due to dummy fiber excitation [5.7]

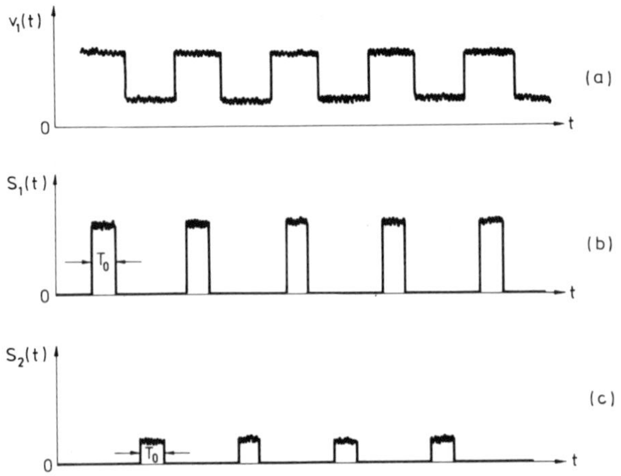

Fig. 5.5a-c. Waveforms used in the spectral-radiometer

The output power from the fiber under measurement is detected by a germanium diode, and amplified. The output at this point appears as in Fig.5.5a. (For the sake of clarity the noise has been kept low whereas, in the worst case, the signal may in fact be buried deep in the noise). This signal is fed to a synchronized "spectral-radiometer", which synchronously generates the signals $S_1(t)$, and $S_2(t)$, as shown in Figs.5.5b and c, respectively. The other function of the spectral-radiometer is to carry out the following mathematical operation

$$v_0(T_{int}) = \int_0^{T_{int}} [S_1(t) - S_2(t)]dt \quad . \tag{5.3}$$

An analog approach for the implementation of this operation is schematically illustrated in Fig.5.6. With this type of circuit, careful design and the use of high quality amplifiers makes it possible to achieve integration times of up to about 15 to 20 minutes.

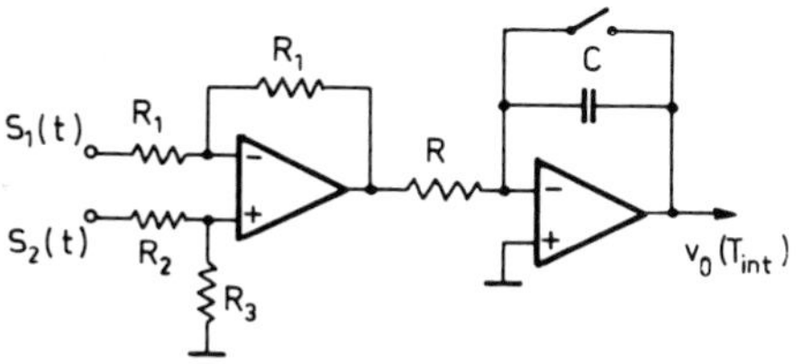

Fig. 5.6. Analog integrating circuit for signal processing in the spectral-radiometer

Both with the use of a spectral-radiometer, and with the use of a lock-in amplifier, the output signal is usually registered by a digital voltmeter, whose output can be interfaced to a computer. Thus, even with the use of a lock-in amplifier, some form of data averaging (equivalent to integration) can be performed. With either type of set-up, automation can be easily achieved by stepping the monochromator with a motor.

Next we consider splice loss, which in principle can be measured in the same way as fiber loss. The actual method depends on the accuracy of the measurement equipment and on the loss level of the splice. Thus, with equipment whose error is much less than the splice loss, we only need to measure the relative change in the output power of a long fiber with and without the splice. If, on the other hand, the splice loss is of the order of the equipment errors, other methods must be used. One approach is to measure a long cascade of splices, and to calculate the mean loss.

We now turn to the measurement of length as dictated by (5.1). As mentioned previously, a convenient means is to use the OTDR, which is based on the observation of the time interval between front and back end Fresnel reflections from the fiber. The principle of this measurement can be seen in Fig.5.7. A short optical pulse (generated by a laser diode or by some other source such as a Nd-YAG laser) is coupled into the fiber using a directional coupler, such as a beam splitter or a fiber based device. The orientation of the directional coupler is such as to guide the reflections onto the detector, as schematically shown in Fig.5.7. To obtain a better reflection from the back end of the fiber, the end can be butted against a front reflecting mirror. The time interval can be determined either directly, using a time interval counter, or indirectly, using a sampling oscilloscope (Fig.5.7). With the latter, the delay generator can be used to scan from the front reflection to the back reflection. The time difference is then the difference between the initial and final settings of the delay generator, for coincident positions of the front and back reflections. The delay generator must of course have sufficient stability, resolution, and accuracy.

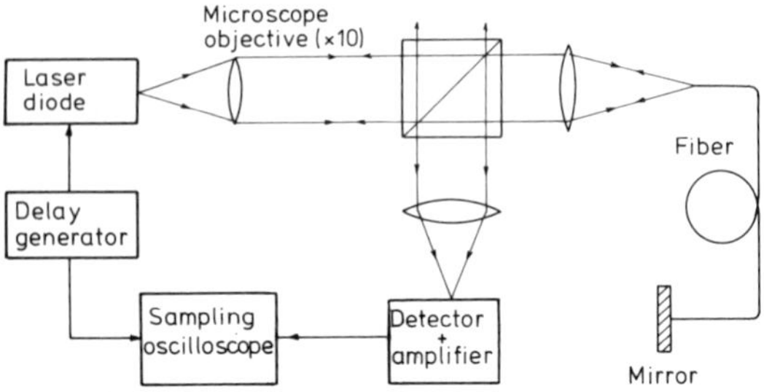

Fig. 5.7. Principle of the optical time domain reflectometer (OTDR)

To obtain the length of the fiber, we must know the velocity of light in the fiber. This can either be estimated from the refractive index, or measured by using a known short length of the same fiber. In the latter case, the length is simply given by

$$L = \frac{\Delta\tau}{\Delta\tau_0} \cdot L_0 \quad , \tag{5.4}$$

where L_0 is the known fiber length, while $\Delta\tau$ and $\Delta\tau_0$ are the time intervals (between front and back reflections) for fiber lengths L and L_0, respectively. For example, if L_0 is 5 m (± 1 cm), $\Delta\tau = 10\ \mu s$, $\Delta\tau_0 = 50$ ns, and the timing accuracy is ± 1 ns, then the error in L is mostly due to the error in the determination of $\Delta\tau_0$, so that we approximately obtain $L = 1000 \pm 20$ m.

The described set-up also has another interesting feature [5.8,9]. Suppose that a short pulse of peak power P_0 is launched into the core of a fiber (Fig.5.8). For the sake of simplicity, we will also assume that the pulse excites all core modes, and that under these conditions the average power attenuation coefficient is a fraction α_1 (per unit length). Suppose further that the power incident at a distance z along the fiber is P(z), and an amount of power ΔP is lost in an infinitesimally narrow segment Δz (Fig.5.8). We can then write the rate equation ($\Delta z \to 0$)

$$\frac{dP(z)}{dz} = -\alpha_1 P(z) \quad , \tag{5.5}$$

whose solution is simply

$$P(z) = P_0 \exp(-\alpha_1 z) \quad . \tag{5.6}$$

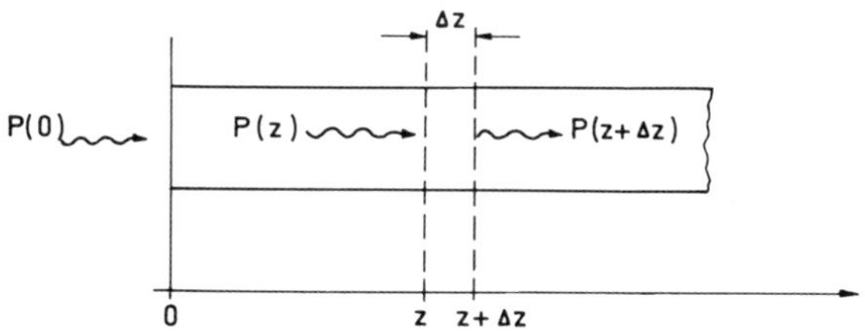

Fig. 5.8. Geometrical interpretation of axial power flow

Let us assume that a fraction S of this power is Rayleigh scattered and captured for backward propagation. This power, say P_r, will arrive back at z = 0, after having been further attenuated by $\exp(-\alpha_2 z)$, in the same manner as in (5.6). (Note that the forward and backward attenuation coefficients are generally different). Thus, the power seen at z = 0 due to excitation by P(z) at (z) is

$$P_r = SP_0 \exp[-(\alpha_1 + \alpha_2)z] \quad . \tag{5.7}$$

If we use the launching time as reference (t = 0), then this power will reach the front face at a time t = 2z/v, where v is the velocity in the fiber. Replacing z in (5.7) by vt/2, we obtain that the temporal response of the backscattered radiation observed at the input is

$$P(t) = SP_0 \exp(-\bar{\alpha} vt) \quad , \tag{5.8}$$

where $\bar{\alpha}$ is the arithmetic mean of α_1 and α_2. Thus, if the amplification of the detector-amplifier unit in Fig.5.7 would be made sufficiently high, a signal such as P(t) in (5.8) would be observed. However, we must remember that under our present conditions, the Fresnel reflection from the front face of the fiber ($\sim$4% of incident power) will be large compared to the back-scatter. To obtain a quantitative estimate, we split S in (5.8) into two components: the fraction of power scattered (say α_s per unit length), and the backward coupling efficiency γ_s. If we then assume a pulse width of T_0, which corresponds to a propagation length vT_0 within which Rayleigh scattering is approximately constant, we can then rewrite (5.8) in the following form

$$P(t) = \alpha_s \gamma_s v T_0 P_0 \exp(-\bar{\alpha} v t) \quad . \tag{5.9}$$

For isotropic scattering, γ_s is approximately given by [5.8]

$$\gamma_s = (NA)^2/(4n^2) \quad , \tag{5.10}$$

where (NA) is the numerical aperture, and n is the approximate refractive index of the core ($\sim$1.5). As an example, let us assume a Fresnel reflection of 0.04 P_0, an NA of 0.2, a pulse width of 30 ns, a Rayleigh loss of 3 dB/km corresponding to an α_s of about 5×10^{-4} m^{-1} at the appropriate wavelength, and a waveguide velocity of 2×10^8 m/s. A quick calculation then shows that the scattered signal immediately after the front face reflection is approximately 35 dB below the Fresnel reflection.

This large difference between the levels places a severe demand on detector-amplifier design. Usually, the problem is side-stepped by index matching the fiber input, by tilting the front face (to misdirect the Fresnel reflection), or by polarizing the launch radiation and observing with a crossed analyzer. Since scattered radiation is randomly polarized, it can pass through the analyzer while the first reflection is blocked. Another alternative, which is used mostly with photomultiplier tubes is to gate the detector and make it insensitive at the moment of reflection.

Let us now refer to (5.9) and observe the following. The power $P(t_1)$ received at time t_1 is a direct measure of the power received from a point z_1 along the fiber. Similarly, $P(t_2)$ is the power received from a point z_2. Any observed difference between $P(t_1)$ and $P(t_2)$ must be due to the two-way loss between z_1 and z_2. Thus, the mean attenuation in the section $|z_1 - z_2|$ must be given by

$$A = \frac{10\ \lg[P(t_1)/P(t_2)]}{2|z_1 - z_2|}\ [\text{dB/unit length}] \quad . \tag{5.11}$$

From (5.9), we know that, in any fiber section with constant attenuation per unit length, a plot of 10 lg[P(t)] against 2z (or vt) must be a straight line. Thus, over any straight line section in such a plot, the attenuation of the fiber is given by the slope of the line (in dB's per unit length). Similarly, the attenuation due to discontinuities (such as splices) can be directly read.

The actual measurement can be performed by using the same equipment as that used for length measurement (Fig.5.7), but excluding the mirror. As already mentioned, some precautions must be taken to prevent detector-amplifier saturation. In addition, because of the inherently low signal-to-noise ratio, some form of signal processing is usually necessary. With the apparatus of Fig.5.7, a convenient means is to integrate the y-output (slow) of the sampling oscilloscope, before acquisition by a plotter or a computer. Alternatively, the two operations of sampling and integration can be combined in a box-car integrator. For convenience, a logarithmic amplifier can also be used to eliminate logarithmic plotting. A practical plot using this technique is shown in Fig.5.9, where splice attenuations can also be seen. Finally,

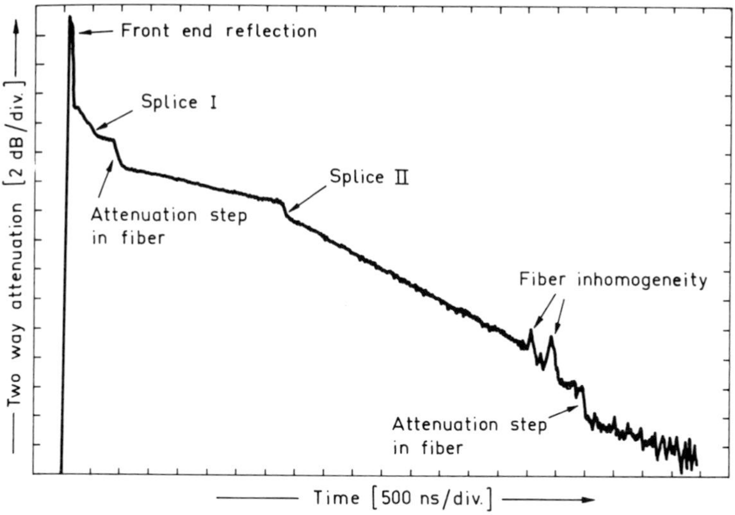

Fig. 5.9. Backscatter waveform from spliced fiber (by courtesy of M. Pikkarainen, General Direction of Posts and Telecommunications, Helsinki, Finland)

we observe the interesting feature that only one end of the fiber is required with the backscatter method.

In concluding this section, we note that for many applications it is sufficient to know the attenuation of the fiber link at a single wavelength. In such cases, it is possible to launch a substantial amount of light power by, for example, using a laser diode. Under such conditions, the apparatus is considerably simpler, since the lamp, monochromator, chopper, and lenses can be completely eliminated. In these cases, a step-index fiber "pig-tail" can be directly bonded to the laser diode (or LED), and the other end of the pig-tail can be butted against the fiber being measured. Similarly, the output end of the fiber can also be simply butted against the detector. Any simplification in the electronics depends mostly on the attenuations to be measured. For large signal-to-noise ratios (low attenuations), relatively simple electronics suffice, whereas poor signal-to-noise ratios require the use of our earlier considerations.

5.3 Impulse Response

In Sect.5.1 we discussed the general difficulties associated with both impulse and frequency response measurements. In this section we will simply assume that the fiber has been fully excited in some manner, for example, using a step-index dummy fiber, as used for attenuation measurements [5.7]. This approach leads to repeatable measurements and allows us to concentrate on the technique itself.

The basic measurement set-up is shown in Fig.5.10. A short optical pulse, which contains spectral (base-band) components at least up to the maximum frequency of interest, is injected into the fiber. Such pulses may, for example, be generated by means of GaAs laser diodes or mode-locked lasers.

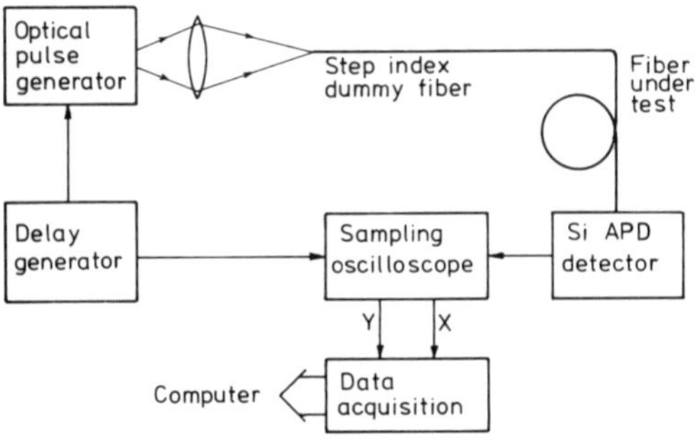

Fig. 5.10. Essential features of the impulse response measurement set-up

The output is usually detected by fast avalanche photodiodes and displayed on a sampling oscilloscope. The displayed waveform is then slowly swept out, converted into digital form, and fed into a computer. Due to dispersion in the fiber, the output pulse is broadened and distorted. Of course, the pulse deformation is partly due to the measurement apparatus. Moreover, the input pulse has its own shape, which is certainly not an impulse. To account for these two factors, we must also record the input pulse shape either directly or by using a short piece of fiber. Note also that the pulse repetition rate must be sufficiently slow to avoid any overlapping tails from neighbouring pulses. With this proviso, we can effectively ignore the periodicity of the signal and only consider individual pulses. Suppose then that the input optical pulse shape is given by a function $S(t)$, and that the impulse response of the dummy fiber, the fiber being measured, the detector, and the oscilloscope are given by $h_1(t)$, $h(t)$, $h_2(t)$, and $h_3(t)$, respectively. The displayed pulse is then given by [5.10]

$$v_1(t) = S(t) * h_1(t) * h(t) * h_2(t) * h_3(t) \quad , \tag{5.12}$$

where the symbol (*) represents convolution. Similarly, if we observe the output of the assumed dummy fiber, the displayed signal is given by

$$v_2(t) = S(t) * h_1(t) * h_2(t) * h_3(t) \quad . \tag{5.13}$$

In order to eliminate the influence of the measurement set-up, we take Fourier transforms of (5.12) and (5.13) and divide. Convolution in the time domain transforms to multiplication in the frequency domain, and we are left with

$$\frac{F[v_1(t)]}{F[v_2(t)]} = F[h(t)] = H(j\omega) \quad , \tag{5.14}$$

where F represents Fourier transformation, and $H(j\omega)$ is the frequency response of the fiber. To obtain the impulse response, an inverse Fourier transformation is necessary, and we have

$$h(t) = F^{-1}\left\{\frac{F[v_1(t)]}{F[v_2(t)]}\right\} \quad . \tag{5.15}$$

Thus, the measurement procedure is to acquire $v_1(t)$ and $v_2(t)$, feed their pulse shape data into a computer, and take Fourier transforms using standard Fast Fourier Transform (FFT) techniques. Division then yields the frequency response of the fiber. As this is often sufficient, the second step of inverse transformation is not always needed.

In a practical measurement, oscilloscope and detector noise may sometimes be a problem so that some form of signal processing may be desirable before computation. Perhaps the most convenient method is to use the computer itself for averaging data samples, this being equivalent to integration. In terms of practical details, the main problem may be related to the generation and detection of short optical pulses. Mode-locked laser sources (e.g. Nd-Yag) with pulse widths of a few hundred picoseconds are convenient, but tend to be expensive and bulky, particularly for field use. An alternative is to use GaAs diodes with abrupt current-power characteristics. Lasers of this type exhibit strong relaxation oscillation (Sect.4.2), and when driven by current pulses of suitable duration and amplitude, the first over-shoot can be extracted [5.11]. A convenient means for generating the required current pulses is to discharge capacitors by avalanche switching [5.12]. The detection of such short pulses is usually achieved by means of avalanche photodiodes. The diodes can be mounted in a strip-line or co-axial circuit, in order to improve high speed performance.

Two further points need discussion. The first is related to the nonlinear response of avalanche photodiodes, because of which it is essential that the output of the dummy fiber be attenuated to a level similar to that found at the output of the long fiber. Over a small range of signals, the detector is essentially linear and (5.12) and (5.13) remain valid. The second point is related to the applicability of linear transforms to the fiber. We must bear in mind that, at sufficiently high power densities in the fiber (e.g. when using Nd-Yag lasers), nonlinear effects such as Raman scattering can occur and the described approach (by Fourier transformation) loses validity. However, the fiber can be considered linear [5.13] at low signal levels. Hence, high power sources must be sufficiently attenuated to ensure linear operation.

5.4 Frequency Response

Measurement of the frequency response is an alternative technique for estimating the dispersion in a fiber. If a sinusoidally modulated optical signal is injected into a fiber, the amplitude of the signal at the output of the fiber is determined by the amplitude response of the fiber, while the phase is shifted by the fiber's phase response. The frequency response can be written in polar form as follows:

$$H(j\omega) = |H(\omega)| \exp\{-j[\theta(\omega) + \omega\tau]\} \quad , \tag{5.16}$$

where $|H(\omega)|$ represents the amplitude response, and $[\theta(\omega)+\omega\tau]$ the phase shift. The latter includes a phase shift $\omega\tau$ due to a common propagation delay τ assumed to be independent of ω. Consider, for example, a one-kilometer length of fiber whose core refractive index is about 1.5. The delay in such a fiber is about 5 μs corresponding to a phase shift of $10^5\pi$ radians at 1 GHz. On the other hand, $\theta(\omega)$ is a non-linear deviation in the order of 2π and accounts for fiber dispersion. From this example, it is obvious that the accuracy of phase measurement must also be high to obtain good results. It is equally obvious that the larger is the bandwidth of the fiber under measurement, the more severe is the accuracy requirement. This actually represents one of the main difficulties of direct frequency response determination.

The conceptually straightforward principle of direct frequency response measurement [5.14-16] is shown in Fig.5.11. An optical source (such as a light-emitting-diode, a laser diode or an externally modulated laser) is modulated by a sinusoidal signal from a sweep generator, and the modulated light is injected into the fiber. (The sweep generator is usually included in the network analyzer). The output of the fiber is detected by using, for example, avalanche or PIN photodiodes, and the signal is fed to the second port of the network analyzer. The network analyzer sweeps the frequency range and provides the transfer function $H_1(j\omega)$. The measurement is repeated for a short length of fiber producing a second transfer function, $H_2(j\omega)$. The transfer function $H_1(j\omega)$ includes the transfer functions $H(j\omega)$, and $H_s(j\omega)$, of the fiber and measurement system, respectively, and can be written in the form

$$H_1(j\omega) = |H(\omega)| \cdot |H_s(\omega)| \exp\{-j[\phi(\omega) + \theta_0(\omega)]\} \quad . \tag{5.17}$$

where $\phi(\omega)$ is the phase shift due to the fiber and $\theta_0(\omega)$ is the phase shift due to the measurement system. Similarly, the response obtained using the

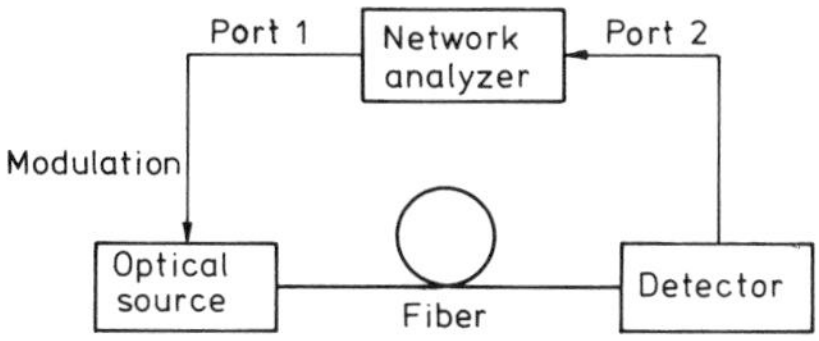

Fig. 5.11. Essential features of the frequency response measurement set-up

short fiber can be written as (neglecting the influence of the short fiber):

$$H_2(j\omega) = |H_s(\omega)| \exp[-j\theta_0(\omega)] \quad . \tag{5.18}$$

It can be seen that division of (5.17) by (5.18) yields the transfer function of the fiber. Thus, in practice, we simply divide the measured amplitude responses and take the difference of the measured phase shifts. Alternatively, when sufficiently well matched detectors are available, a reference signal [corresponding to $H_2(j\omega)$] can be generated by using a beam splitter in the input optical path. In this case, the transfer function is obtained directly. Another variation is to use a spectrum analyzer to obtain the amplitude response and to use a vector voltmeter to measure the phase shifts.

We mentioned earlier the difficulty of accurate phase measurements, and will now discuss an interesting technique proposed by NICOLAISEN and RAMSKOV-HANSEN [5.17] which alleviates it. The principle of the method is shown in Fig.5.12, in which the optical source is driven by the output of mixer 1, which in turn is driven by generators 1 and 2. The mixer is assumed to be a perfect multiplier that produces sum and difference frequencies at its output. Using the notations in Fig.5.12, and labelling $(\omega_1 - \omega_2) = \omega_d$, $(\omega_1 + \omega_2) = \omega_s$, $(\phi_1 - \phi_2) = \phi_d$, and $(\theta_1 + \theta_2) = \theta_s$, it is easy to show that the optical power into the fiber takes the form

$$P_i(t) = A \cos(\omega_d t + \phi_d) + A \cos(\omega_s t + \phi_s) \quad , \tag{5.19}$$

where $A = \eta_1 A_1 A_2/2$ and η_1 is the electro-optical conversion factor. We have also assumed that ω_d and ω_s are nearly equal, so that η_1 is the same at both frequencies. Each frequency component in (5.19) propagates through the fiber, and by the time it reaches the output, its amplitude is multiplied by the

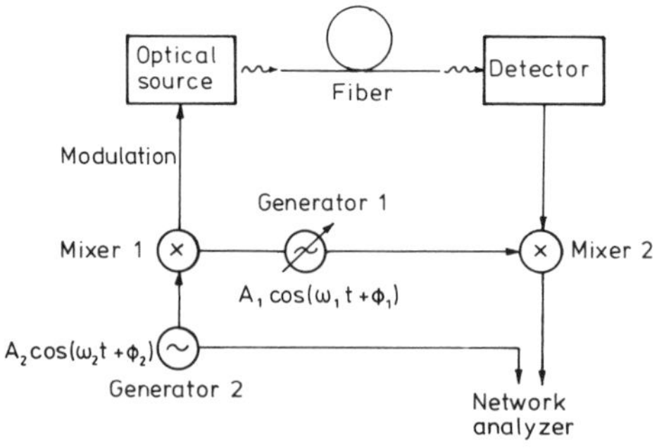

Fig. 5.12. A differential phase measurement method

amplitude response of the fiber [see e.g. (5.16)] while the phase is shifted by the phase shift. The output is then of the form

$$P_0(t) = A|H(\omega_d)| \cos[\omega_d t + \phi_d + \theta(\omega_d) + \omega_d \tau] + A|H(\omega_s)| \cos[\omega_s t + \phi_s + \theta(\omega_s) + \omega_s \tau] \quad . \tag{5.20}$$

This signal is converted into electrical form and fed to mixer 2, where it is multiplied by the output of generator 1. The output of the mixer then has frequency components at ω_2, $(2\omega_1 + \omega_2)$, and $(2\omega_1 - \omega_2)$. If we assume that the latter two components can be filtered out, only the ω_2 component is left. Furthermore, if we assume that the amplitude response varies slowly and that $\omega_d \approx \omega_s$, then

$$|H(\omega_d)| \approx |H(\omega_s)| \approx \frac{|H(\omega_d)| + |H(\omega_s)|}{2} \quad . \tag{5.21}$$

Under these conditions, the output of mixer 2 can be simplified to read

$$v_0(t) = D \cos\left[\frac{\theta(\omega_d) + \theta(\omega_s)}{2} + \omega_1 \tau\right] \cos\left[\omega_2 t + \phi_2 + \frac{\theta(\omega_s) - \theta(\omega_d)}{2} + \omega_2 \tau\right]$$

where

$$D = (A_1^2 A_2 \eta_1 \eta_2 / 4)\,[|H(\omega_d)| + |H(\omega_s)|] \quad , \tag{5.22}$$

and η_2 is the opto-electronic conversion factor.

If we now choose the frequency of measurement such that

$$\omega_1 \tau = n\pi - \frac{\theta(\omega_d) + \theta(\omega_s)}{2} \quad , \tag{5.23}$$

the power detected by the network analyzer is maximized because the cosine term in the amplitude of (5.22) becomes ± 1. The network analyzer compares the resultant signal to $A_2 \cos(\omega_2 t + \phi_2)$ and displays the following amplitude and phase responses:

$$D/A_2 = \frac{\eta_1 \eta_2 A_1^2}{4} [|H(\omega_d)| + |H(\omega_s)|] \approx \frac{\eta_1 \eta_2 A_1^2}{2} \cdot |H(\omega_1)| \quad , \tag{5.24}$$

and

$$\Delta\phi = \frac{\theta(\omega_s) - \theta(\omega_d)}{2} + \omega_2 \tau \quad . \tag{5.25}$$

It can be seen that (5.24) gives the fiber response at the frequency ω_1 which is selected according to (5.23), while (5.25) gives the differential phase plus the constant term $\omega_2\tau$. Here, the large phase shift problem can be alleviated by selecting a small ω_2. For example, if we select ω_2 = 10 MHz, then $\omega_2\tau = 10^2\pi$. Moreover, the accuracy can be improved by the use of a stable (crystal controlled) oscillator.

We conclude this section by commenting on the equivalence of frequency and time domain measurements. This is easily understood if we remember, for example, that the frequency spectrum of a Gaussian pulse of width T is also Gaussian and has a spectral width of about 1/T. Thus, an optical pulse which is 200 ps wide (e.g. from a Nd-Yag mode locked laser) contains spectral components up to about 5 GHz. If the pulse would be sharper, instead of Gaussian, then spectral components at even higher frequencies would be found. This feature makes it possible to directly measure the frequency response of a fiber that is excited by pulses [5.18]. Often, it is more convenient to perform a time domain measurement as described in Sect.5.3, in which case the phase information is automatically available. The disadvantage is that a computer must be used to carry out the Fast Fourier Transforms.

5.5 Refractive Index Profile

The refractive index profile of a fiber has a decisive effect on its bandwidth and therefore on its information carrying capacity. As a result, profile measurement is of interest both for improving the understanding of propagation behaviour in a fiber, and for profile control during production. Several methods are presently used, amongst them interference microscopy [5.19], near-field scanning [5.20], the scattering method [5.21], the reflection method [5.22,23], spatial filtering [5.24], the refracted near field technique [5.25], as well as many variations and refinements [5.26-29]. All of them have their merits, but we choose to discuss the near field scanning method because it has the advantage of simplicity and it is fast. Moreover, the discussion of the theory behind the method provides useful insight into fiber behaviour.

Before discussing the method itself, let us return to Sect.3.6 and refer to Fig.3.41, where a geometrical interpretation of the propagation vectors is given. Strictly, such a geometrical optics approach is only valid for vanishingly small wavelengths or infinitely large fiber dimensions. We will restrict ourselves to multimode fibers which typically have diameters of about 50 λ. Defining the angle between $k(\rho)$ and β to be $\theta_f(\rho)$ we can write

$$\cos\theta_f(\rho) = \beta/k(\rho) \quad . \tag{5.26}$$

As we saw in Sects.3.5 and 3.6, minimum β corresponds to the refractive index of the cladding, and is given by $\beta_c = 2\pi n_c/\lambda$. Thus, the maximum angle that the ray can make is

$$\cos\theta_c(\rho) = \frac{n_c}{n(\rho)} \quad . \tag{5.27}$$

At the input of the fiber, this corresponds to a ray incident at an angle $\theta_0(\rho)$, related to $\theta_c(\rho)$ by Snell's Law. Assuming an air-fiber interface, we obtain

$$A(\rho) = \sin\theta_0(\rho) = n(\rho)\,\sin\theta_c(\rho) = [n^2(\rho) - n_c^2]^{\frac{1}{2}} \tag{5.28}$$

Only the radiation incident into the cone defined by $\theta_0(\rho)$ can propagate in the fiber. Let us suppose that the end of the fiber is illuminated by a Lambertian source, which emits uniformly across its area. Then the power accepted into the cone is given by (see e.g. [5.1])

$$P\,(\rho) = A_s \int_0^{2\pi} d\phi \int_0^{\theta_0(\rho)} B(\theta)\,\sin\theta\,d\theta \quad , \tag{5.29}$$

where

$$B(\theta) = B\,\cos\theta \tag{5.30}$$

is the brightness of the Lambertian source, A_s is the source area, and ϕ varies over 2π in the fiber plane. Performing the simple integration, we find that the power is directly proportional to the numerical aperture. Thus, we can write

$$\frac{P(\rho)}{P(0)} = \frac{A^2(\rho)}{A^2(0)} = \frac{n^2(\rho) - n_c^2}{n^2(0) - n_c^2} \quad . \tag{5.31}$$

If all modes propagate without differential attenuation or mode conversion (generally implying a short fiber), then the same power distribution should be found at the output. For the α-profile representation of (3.111), (5.31) reduces to

$$\frac{P(\rho)}{P(0)} = 1 - (\rho/a)^{\alpha} \quad , \tag{5.32}$$

or

$$\lg[1 - P(\rho)/P(0)] = \alpha \lg(\rho/a) \quad . \tag{5.33}$$

Hence, a lg-lg plot of $[1 - P(\rho)/P(0)]$ against (ρ/a) yields a straight line of slope α.

In cases involving leaky modes with low attenuation, the measured power profile will not follow (5.32) exactly, because we have implicitly assumed the contribution of propagating modes only. In such cases, (5.29) must be modified to include angles greater than critical. The output power is then given by

$$P(\rho) = 4A_s \int_0^{\pi/2} d\phi \int_0^{\pi/2} B \cos\theta \sin\theta \exp[-\alpha(\theta,\phi)z] \cdot d\theta \quad . \tag{5.34}$$

We have now included the attenuation term $\alpha(\theta_0,\phi)$ to account for power loss from the leaky modes. This term is zero for propagating modes and infinite for refracted waves. Thus, we can separate the above integral by changing the limits of integration [5.30], and obtain:

$$\frac{P(\rho)}{P(0)} = \frac{A^2(\rho)}{A^2(0)} + \frac{4}{\pi A^2(0)} \int_0^{\pi/2} d\phi \int_{\theta_0(\rho)}^{\pi/2} \cos\theta \sin\theta \cdot \exp[-\alpha(\theta,\phi)z] \cdot d\theta$$

$$= \frac{n^2(\rho) - n_c^2}{n^2(0) - n_c^2} \cdot C(\rho,z) \quad , \tag{5.35}$$

where

$$C(\rho,z) = 1 + \frac{4}{\pi[n^2(\rho) - n_c^2]} \int_0^{\pi/2} d\phi \int_{\theta_0(\rho)}^{\pi/2} \cos\theta \sin\theta \exp[-\alpha(\theta,\phi)z] \cdot d\theta \quad . \tag{5.36}$$

We see that the measured power profile must be divided by the correction factor $C(\rho,z)$. In order to evaluate this factor, we first cast it into modal form by using the geometrical intepretation afforded by the WKB method (Fig. 3.41). Moreover, we must bear in mind that the above equations apply externally to the fiber, and we must apply Snell's law to change θ to the equivalent internal angle θ_f. Consequently, we use the following changes of variables

$$\left.\begin{aligned} \sin\theta &= n(\rho) \sin\theta_f \quad , \quad \text{or} \\ \cos\theta \cdot d\theta &= n(\rho) \cos\theta_f \cdot d\theta_f \quad , \end{aligned}\right\} \tag{5.37}$$

$$\left.\begin{aligned} \cos\theta_f &= \beta/k(\rho) \quad , \\ \sin\phi &= (\nu/\rho)/\sqrt{k^2(\rho) - \beta^2} \quad , \end{aligned}\right\} \tag{5.38}$$

where ϕ is the angle between the projection of $k(\rho)$ onto the fiber plane and the radial component of $k(\rho)$, while θ_f is the angle between $k(\rho)$ and β. The correction factor then takes the form

$$C(\rho,z) = 1 + \frac{4}{\pi A^2(\rho)} \int_0^V \frac{d\nu}{a^2\rho k_0^3} \int_{u_c(\rho)}^{\sqrt{V^2+\nu^2}} \frac{\{\exp[-\alpha(u,\nu)\cdot z]\}u\cdot du}{\sqrt{n^2(\rho) - n^2(0) + \dfrac{u^2}{k_0^2a^2} - \dfrac{\nu^2}{k_0^2\rho^2}}} \ , \tag{5.39}$$

where

$$u_c(\rho) = \sqrt{V^2(\rho/a)^\alpha + (a/\rho)^2\nu^2} \quad \text{for} \quad \rho > 0 \quad , \quad V < u < \sqrt{V^2 + \nu^2} \quad . \tag{5.40}$$

In the above $k_0 = 2\pi/\lambda$, $u^2 = a^2[k_0^2n^2(0) - \beta^2]$, and $V^2 = a^2k_0^2[n^2(0) - n_c^2]$. The limits of integration are obtained as follows. If we refer to Fig.3.37b, we see that the leaky mode caustics at ρ_2 and ρ_3 will approach each other as we go deeper into the leaky mode regime, and eventually meet at $\rho = a$ when the mode starts to be refractive. This situation is shown in Fig.5.13, from which we conclude that at this limit

$$\beta^2 = k_0^2n_c^2 - \frac{\nu^2}{a^2} \quad , \tag{5.41a}$$

corresponding to an upper limit on u given by

$$u_{max} = \sqrt{V^2 + \nu^2} \quad . \tag{5.41b}$$

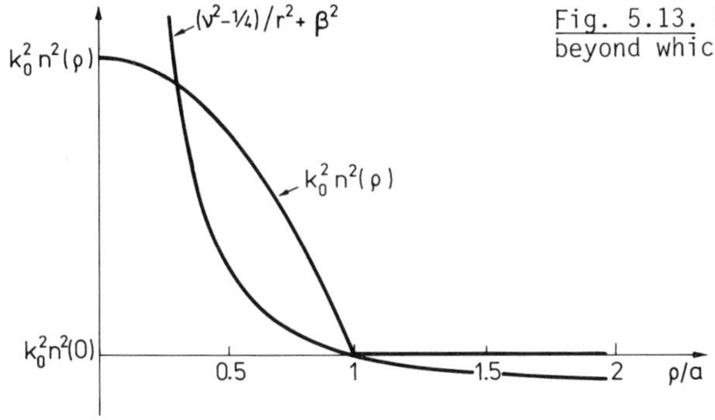

Fig. 5.13. Limit of leaky mode region beyond which the mode turns refractive

The lower limit on u can be similarly obtained by inspection and turns out to be the zero of the square root term. Moreover, the upper limit on ν is approximately V (near-parabolic profile), and leaky modes can exist for all ν values up to V. Consequently, we must tailor $u_c(\rho)$ for each ν value over which we integrate. The minimum value for $u_c(\rho)$ is obviously V (Figs.3.38 and 39), but even for small values of ν, $u_c(\rho)$ can become significantly greater than V, at least at small values of (ρ/a). Hence, for numerical evaluation, we first choose the value of ρ at which the correction factor is required. Then, for each allowed value of ν, we evaluate $u_c(\rho)$. If the result is less than V, then we set it equal to V; however, if it is greater than V, then the calculated value is used. (The maximum allowed value must, of course, be limited to the upper limit on u, since this represents the beginning of refraction). We then integrate over u and repeat the procedure for the other values of ν.

Before we can perform this integration, we need to evaluate the leaky mode attenuation coefficient $\alpha(u,\nu)$. From Sect.3.6, we know that the field decays exponentially beyond the turning point [see (3.118)]. Therefore, for unit incident power at the inner caustic (ρ_2), the fractional power left at the outer caustic (ρ_3) is given by

$$T = \frac{|E(\rho_3)|^2}{|E(\rho_2)|^2} = \exp\left[-2\int_{\rho_2}^{\rho_3}\sqrt{\frac{\nu^2}{\rho^2} - k^2(\rho) + \beta^2}\,d\rho\right] \quad . \tag{5.42}$$

Furthermore, if Δz is the distance between two "reflections", the length-dependent leaky-mode attenuation factor becomes

$$\alpha(u,\nu) = T/\Delta z \quad . \tag{5.43}$$

The distance Δz is simply obtained from the ray equation

$$\Delta z = 2\int_{\rho_1}^{\rho_2}\left[\frac{dz}{d\rho}\right]d\rho \tag{5.44}$$

by once again using the geometrical interpretation of the WKB method. Some simple manipulation then yields

$$\Delta z = 2\beta\int_{\rho_1}^{\rho_2}\frac{d\rho}{\sqrt{k^2(\rho) - \beta^2 - \frac{\nu^2}{\rho^2}}} \quad . \tag{5.45}$$

Hence, (5.43) can be solved via (5.42) and (5.45). The result can then be substituted into (5.39) to obtain the correction factor. The obvious problem is that we are trying to correct a measurement of the refractive index profile of which we require prior knowledge in order to perform the correction. One approach would be to use an iterative procedure starting from the uncorrected power profile, but this would obviously be rather tedious and time consuming. Alternatively, we could adopt the approach of ADAMS et al. [5.31] and assume that the correction factors for near-parabolic profiles will not vary much from those calculated for the perfectly parabolic variation. Under these conditions, (5.42) and (5.45) can be solved analytically as follows. First, we assume a refractive index profile of the form

$$\left.\begin{aligned} n^2(\rho) &= n^2(0)[1 - 2\Delta(\rho/a)^2] \qquad r \leqq a \\ &= n^2(0)[1 - 2\Delta] \quad . \qquad r > a \end{aligned}\right\} \tag{5.46}$$

The solution of (5.45) is then simply achieved by substituting (5.46) into (5.45), using a change of variable $x = \rho^2$, and selecting an arcsin solution which is tabulated [5.32]. We obtain the result

$$\Delta z = \frac{\beta a^2}{V^2} \arcsin \left\{ \frac{u^2 - 2V^2\rho^2/a^2}{\sqrt{u^4 - 4V^2\nu^2}} \right\}_{\rho_1}^{\rho_2} \quad , \tag{5.47}$$

where

$$u = a[k_0^2 n^2(0) - \beta^2]^{\frac{1}{2}} \quad , \tag{5.48}$$

$$V = ak_0 n(0)\sqrt{2\Delta} \quad , \tag{5.49}$$

and a weakly guiding fiber has been assumed. Evaluation of (5.47) is conveniently achieved by observing that, at the caustics, we have

$$\beta^2 = k^2(\rho) - \nu^2/\rho^2 \quad , \tag{5.50}$$

so that substitution into (5.48), and solution of the quadratic equation yields the results

$$\frac{\rho_{1,2}^2}{a^2} = \frac{u^2 \pm \sqrt{u^4 - 4V^2\nu^2}}{2V^2} \quad . \tag{5.51}$$

Use of these limits in (5.47) then leads to the simple formula

$$\Delta z = \frac{\beta a^2}{V} \cdot \pi \quad . \tag{5.52}$$

Evaluation of (5.42) proceeds similarly, except that the integral is separated into two parts: one from ρ_2 to a, and the second from a to ρ_3. The following result is obtained

$$T = e^{-2I}$$

where

$$I = \frac{u^2}{2Va}\left(\frac{\pi}{2} - \arcsin \frac{u^2 - 2V^2}{\sqrt{u^4 - 4V^2\nu^2}}\right) + \nu\left[\ln\left|\frac{\nu}{\sqrt{u^2 - V^2}}\left(1 + \sqrt{1 - \frac{u^2 - V^2}{\nu^2}}\right)\right| - \sqrt{1 - \frac{u^2 - V^2}{\nu^2}}\right] \quad . \tag{5.54}$$

This lengthy expression can be further simplified by assuming that the power decay from $\rho = \rho_2$ to $\rho = a$ is negligible, which implies a slowly decaying leaky mode. Instead of integrating from ρ_2 to ρ_3, we then integrate from a to ρ_3, so that the integral is just the second part of (5.54), and some manipulation yields the approximate form used by ADAMS et al. [5.30]

$$T \approx \left[\frac{u^2 - V^2}{(\nu + x)^2}\right]^{\nu} \exp(2x) \quad , \tag{5.55}$$

where

$$x = \sqrt{\nu^2 - u^2 + V^2} \quad . \tag{5.56}$$

We can now compute the necessary near-field correction factor by substituting (5.52) and (5.55) into (5.39). The result of such a computation is shown in Fig.5.14, for an on-axis index of 1.46, a fiber diameter of 50 µm, and a wavelength of 900 nm. It can be seen that the correction factor varies over a significant range for various fiber lengths and V values.

We now turn to the measurement of the power profile at the output end of the fiber. The principle of the measurement is depicted in Fig.5.15. The light from a Lambertian source is chopped, filtered, and launched into the fiber by using a lens having a numerical aperture larger than that of the fiber. The magnified image at the other end of the fiber is focused onto a detector, which can be used to scan the image. As in previous sections, the signal-to-noise ratio of the measurement can be improved by using a lock-in amplifier, or the spectral radiometer of Sect.5.2.

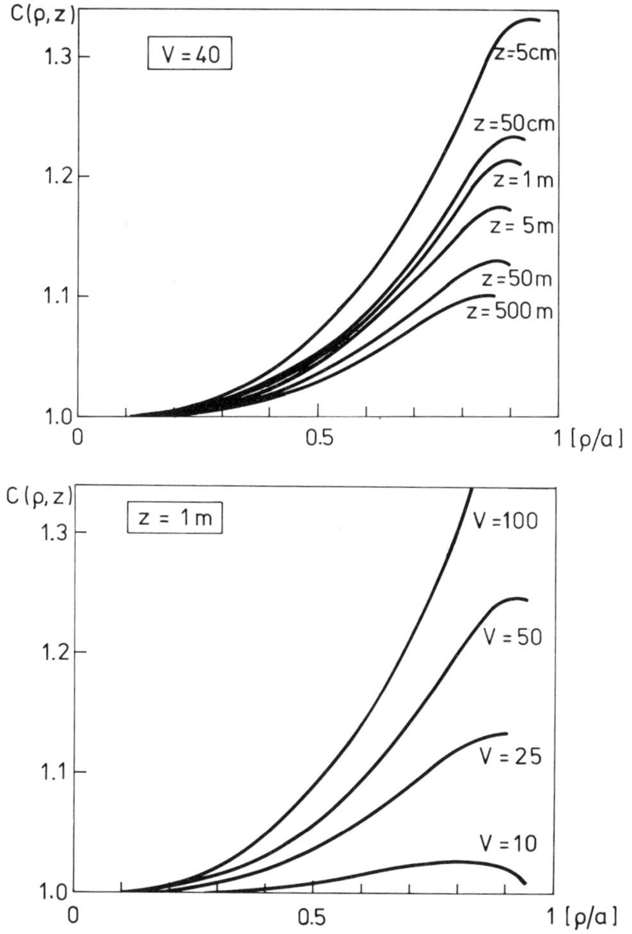

Fig. 5.14. Leaky mode correction factor as a function of normalized radial position: (a) with the fiber length z as parameter, and (b) the V value as parameter

Another interesting variation is to replace the detector by a TV camera that is sensitive in the infra-red region. The video signal generated by the magnified image can then be analyzed to produce the intensity profile. The simplest variation is to observe the required horizontal line (corresponding to the center of the fiber) on an oscilloscope. However, commercial analyzers of varying degrees of refinement are also available to perform this function, and they offer the advantage of ease of coupling to a computer. As before, the computer can then be used for signal averaging. The major disadvantage of this approach is the potentially inhomogeneous response of the TV camera target, which requires point by point calibration but ultimately leads to additional errors.

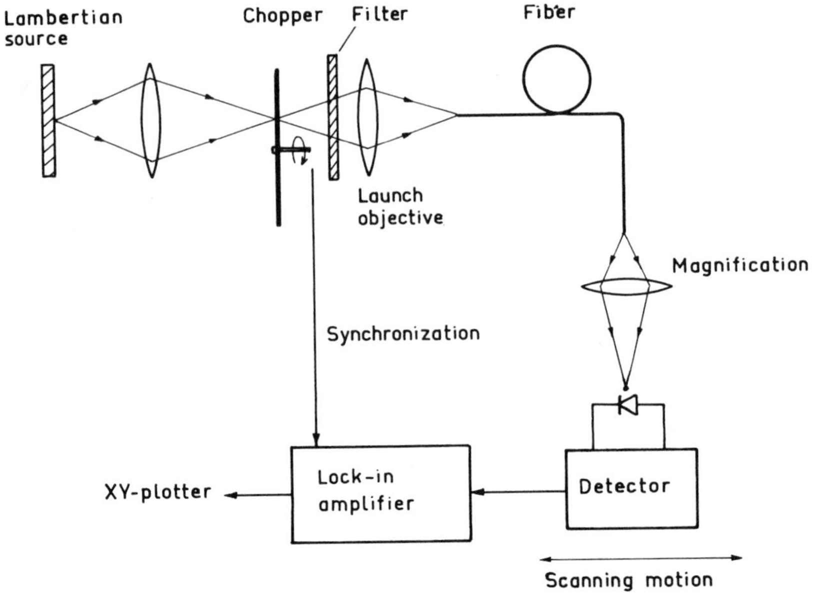

Fig. 5.15. Essential features of the near field measurement set-up

Having measured the intensity profile, we can correct it to obtain the relative refractive index profile. This may be sufficient for many purposes, but generally the terms n(0), Δ, and a of the α-profile [see (3.111)] must also be determined before the complete description is achieved. A convenient means for determing the product $2n(0)\cdot\Delta = A(0)$ is to observe the far-field radiation pattern of the fiber. GLOGE and MARCATILI [5.20] have shown that the far field is approximately given by

$$\frac{P(\theta)}{P(0)} = \left[1 - \frac{\sin^2\theta}{2n^2(0)\Delta}\right]^{2/\alpha} \approx \left[1 - \frac{\sin^2\theta}{A^2(0)}\right]^{2/\alpha} \quad . \tag{5.57}$$

Thus, the sine of the angle at which the output power goes to zero corresponds to the on-axis numerical aperture. In practice, a sharp transition to zero is not always observed and a threshold level (e.g. 10% of peak power) has to be used. The measurement itself is straightforward, and the same equipment as for the near-field measurement can be used. The only difference is that the detector is mounted on a rod ($\sim$50 cm) that is pivoted at a point corresponding to the fiber end and the magnifying lens is not used in this case. If the previously mentioned TV camera arrangement is used, the fiber end can directly illuminate the camera target. In both situations, an accurate measure of the distances is necessary to be able to specify the angles. Finally, we must have knowledge of n(0) to complete the refractive index profile description. For most practical fibers, the maximum index difference is small (~1%) and it is reasonable to assume that n(0) is given by the refractive index of the bulk material before index grading.

In concluding this section, we should like to shortly review an interesting technique for profile determination described by OLSHANSKY and OAKS [5.33].

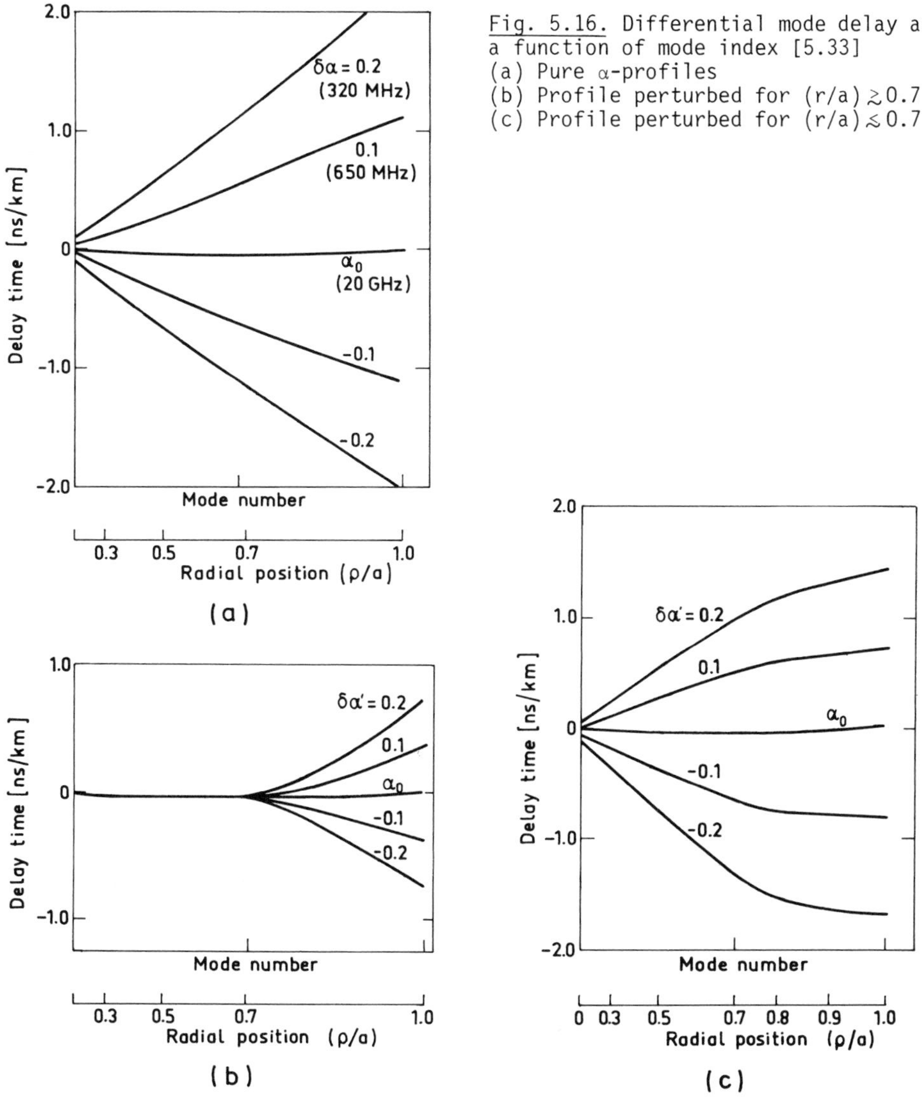

Fig. 5.16. Differential mode delay as a function of mode index [5.33]
(a) Pure α-profiles
(b) Profile perturbed for $(r/a) \gtrsim 0.7$
(c) Profile perturbed for $(r/a) \lesssim 0.7$

The advantage of the method is its high sensitivity to very small changes in the profile, and its capability of showing deviations from pure α-profiles. In the measurement, the equipment described in Sect.5.3 is used to measure the differential delay amongst different mode groups. The procedure is to scan the input of the fiber (across a diameter) with a small spot, and measure the delay as a function of spot position. The index m of the excited group is approximately related to the radial position by

$$[m/M] \approx (\rho/a)^2 \quad , \tag{5.58}$$

where M is the maximum number of modes that the fiber can support. The delay τ_m of each mode group can be calculated for various types of profiles. Examples of three situations are shown in Fig.5.16a,b and c. The first shows the situation for a pure α-profile and it can be seen that the slope of the curves is a measure of the deviation $\delta\alpha$ from the optimum profile, for which all mode groups are delayed equally. Figure 5.16b shows the delays for a profile that is optimum only up to ρ = 0.7a, while Fig.5.16c illustrates the results for a profile which is optimum beyond ρ = 0.7a. In practice we can

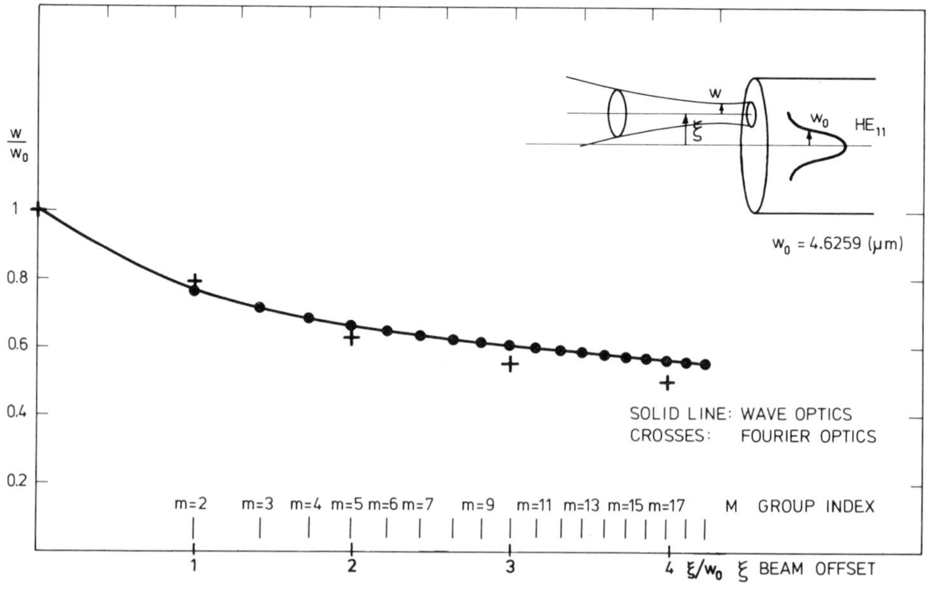

Fig. 5.17. Optimum spot size for selective excitation of a fiber by a Gaussian beam (parabolic refractive index profile)

therefore perform a differential mode delay measurement and, by fitting piecewise linear segments to the curve, we can describe the index variation by a multiple α-profile. To obtain the best resolution, mode conversion must obviously be minimized, while the selective excitation should be optimized. For Gaussian-beam excitation, one might expect that the optimum spot size would correspond to the width of the fundamental mode (HE_{11}) intensity profile in the fiber. However, as shown by SAIJONMAA et al. [5.34], this does not occur, and the spot size must be reduced according to Fig.5.17. For typical multimode fibers, the half width of the fundamental mode field distribution is about 5 μm. Figure 5.17 then indicates that a reasonable compromise would be to use a spot size with a half width of about 3 μm corresponding to an intensity spot size of about 4.2 μm.

6. Fiber Optical Systems and Their Applications

Digital communications systems, based on the simplest modulation method of switching the source on and off, currently represent the largest application area for optical fibers. The basic idea is, of course, rather well known in telegraphy and data transmission (in computer networks), but it can also be used for speech or picture transmission, provided that the analog signal is first sampled and each sample value is subsequently converted into digital form [6.1].

In addition to such systems, other significant application areas for fibers are the analog transmission of analog signals (Sect.6.2), instrumentation and data systems (Sect.6.3), as well as fiber sensors (Sect.6.4). In this chapter, we will discuss these applications in the above order, and for the sake of completeness we will also briefly consider a technique for increasing the information carrying capacity of fibers by using wavelength-division multiplexing techniques.

6.1 Digital Communication Systems

In public telephone networks, the analog speech signals are almost exclusively limited to the frequency range 300-3400 Hz. In spite of this, the digital transmission rates which are required are considerably higher than the ones used in telegraphy and data transmission. However, this disadvantage is amply compensated for by the efficiency with which switching (in exchange) can be realized by using the characteristics of time-division multiplexed channels. In fact, telecommunications administrations and companies are in the midst of a huge conversion process from analog to digital techniques. The ease with which optical sources can be digitally modulated is well suited to these current trends.

In order to understand the framework within which the optical system is required to operate, we should first take a look at the standard digital interfaces. Accordingly, we start from the "speaker's" end, and note that each voice signal is usually sampled at a rate of 8000 Hz, and that each sample is converted into a digital "word", a byte, of 8 bits. The first bit gives the polarity of the signal, while the other seven bits are used to express the magnitude, according to a logarithmic quantization law. By convention, the first three bits (after the polarity bit) are used to express the characteristic, and the last four the mantissa [6.1]. As a result of this procedure, each voice channel requires a transmission rate of 64 k bit/s. For economy, it is usual to time domain multiplex at least 30 channels[1], to form a bit stream rate of 2048 kbit/s (Table 6.1).

Table 6.1. Digital hierarchy according to CEPT and CCITT for synchronous transmission of multiplexed pulse-code modulated voice signals

Level of hierarchy	Digital rate [Mbit/s]	CCITT recommend.	Line code	u [V]	Z_0 [Ω]	No. of voice channels
0	0,064	G.703	-	-	-	1
1	2,048	G.703 G.732 G.734	HDB-3	2.37 3.0	75 120	30
2	8,448	G.703 G.741, 742 G.744	HDB-3	2.37 3.0	75 120	120
3	34,368	G.703 G.751	HDB-3	1.0	75	480
4	139,264	G.703 G.751	CMI	1.0	75	1920
5	565,148					7680

[1] In many countries, such as the USA and Japan, other hierarchies are used, e.g. the first level may have 24 voice channels and a bit rate of 1544 k bit/s. Moreover, the details of multiplex rates, frame properties, and encoding laws are also, in general, different.

In this case, the channel samples are interleaved in time such that each frame consists of 32 consecutive time slots. The 0th time slot is used for synchronization purposes, and the 15th time slot is reserved for the signalling needed for setting up and maintaining the telephone connection. This latter slot is also used for transmitting the dialled number.

As can be seen from Table 6.1, the number of voice channels in hierarchical levels 2...5 consistently increase by factors of 4. To achieve this, the lower-order bit streams are multiplexed bit-by-bit. In addition, special service bits are inserted for frame synchronization, alarm purposes, and for synchronization of the tributaries. This is the reason why the resulting transmission rates are not exact quadruples of the lower order signals.

Whatever the hierarchy, a basic requirement of all these pulse code modulation (PCM) systems is that the transmitted signal should contain sufficient timing (clock) information, to allow synchronous regeneration of the transmitted signal. Such timing information is, of course, inherently contained in the transmitted signal transitions. However, if the signal happens to contain a long sequence of 0's (also 1's in a non-return-to-zero system), then the receiver or regenerator loses synchronization. It is precisely for this reason that line codes, such as the ternary code HDB-3 (high density bipolar), are used. (In wire systems the code is also used to maintain a low average or DC level in the line, in order to avoid useless power dissipation). The encoding law for HDB-3 (the code "CMI" in Table 6.1 will be explained later) is the following:

$$1 \to +1 \text{ or } -1 \text{ alternately}, \quad 0 \to 0 \quad . \tag{6.1}$$

In other words, all logic 1's to be transmitted, are alternately transmitted as positive and negative levels, while zeroes are transmitted as zeroes. In addition, if more than three consecutive zeroes occur, the fourth zero is encoded as a 1 of the same polarity as the previous legitimate 1. These special 1's are known as violations of the general rule, and are interpreted as zeroes by the receiver logic.

Another characteristic of the digital voice link which has been standardized by some authorities, is the allowable error rate. Two different objectives can be defined [6.2]

$$P(\varepsilon) \leqq (20 + 0.5\ L/\text{km}) \cdot 10^{-8} \quad , \quad \text{for} \quad R = 2\ \text{Mbit/s}$$

$$P(\varepsilon) \leqq 5\ L/\text{km} \cdot 10^{-10} \quad , \quad \text{for} \quad R \geqq 8\ \text{Mbit/s} \tag{6.2}$$

where P(ε) is the probability of error and L is the length of the digital transmission section. We should also, of course, adopt these objectives for fiber-optical links such that they are met in the worst case, i.e. at the end of the operational life of the system.

Optical transmission in fiber cables is realized in much the same way as conventional PCM transmission in metallic cables [6.3]. The optical system is connected to one of the PCM interfaces given in Table 6.1, as indicated in Fig.6.1. The interface code HDB-3 must, of course, be translated into a binary code because optical power cannot be negative. Typically, one ternary symbol may correspond to two binary symbols, in which case the modulation rate will be doubled. The binary pulses are then used to turn the optical source (laser or light-emitting diode) on and off. At the receiver, the optical pulses are converted back to electrical form by a semiconductor detector, often together with avalanche multiplication (Chaps.2 and 4).

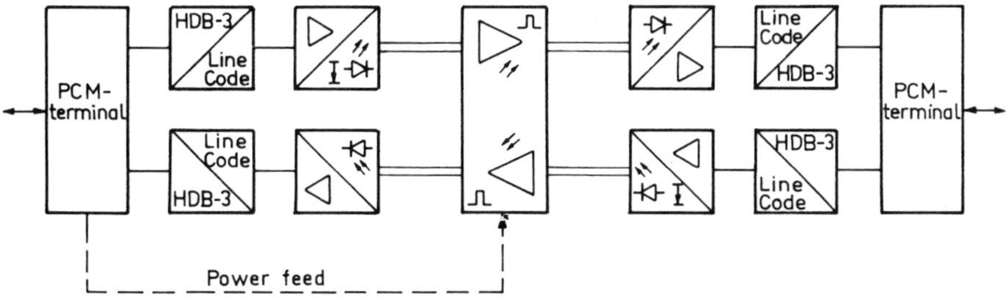

Fig. 6.1. Block diagram of digital system based on optical fibers

For longer PCM links, intermediate repeaters are needed. The pulses are then usually completely regenerated in three distinct steps: amplification and equalization of the pulse waveform, recovery of the clock from the pulse train, and clocked detection and retransmission of the pulses. All this is most conveniently performed electrically by using modifications of conventional PCM circuits. In principle, amplification could also be achieved by using laser amplifiers, but pulse processing by optical means is not presently possible. We should note that the purpose of regeneration is to prevent the gradual deterioration of the signal quality that occurs in analog systems. Obviously, for a long cascade of repeaters, some loss in quality is inevitable due to the accumulation of bit errors and timing jitter.

Before proceeding to fiber systems, we should mention a couple of practical considerations. The first one is related to the fact that, whenever we use repeaters, we must supply power for their operation. In metallic line systems, the required power is simply transmitted along the signal carrying lines in the form of direct current. In optical fiber systems, we could also, in principle, transmit unmodulated optical power, which could then be converted into electrical power by the use of solar cells [6.4]. However, this is not particularly convenient because of the high energy densities that would be involved. Moreover, this technique would feed only two repeaters, one from the transmitter end, and the other from the receiver. (This is mainly due to the inefficiency of "tapping" and converting the optical power). A more practical approach would be to include metallic wire pairs in the fiber cable structure specifically for power feeding purposes. In this case, problems (such as lightening transients, etc.) that are normally associated with metallic pair feed systems could be easily alleviated by filtering and suppression. Power could then be tapped in the conventional way by using zener diodes in series, converters, etc. We should note that these arguments basically apply to junction and trunk lines; however, at the subscriber line level, it may be attractive to use optical power feeding systems [6.5]. We should also observe that fiber systems offer the possibility of long inter-repeater distances; thus, in many cases, local power from the mains could be easily used in conjunction with "back-up" batteries.

The second practical point worth mentioning is that optical transmission becomes more and more economical as the bit rate, and hence the number of voice channels, is increased. This argument can be supported by the following considerations:

1) Low dispersion fibers work equally well at all reasonable bit rates, when the source and its emitted wavelength have been properly chosen.

2) The inter-repeater distance, at a given wavelength, does not depend strongly on the bit rate, provided that the link is limited by attenuation rather than dispersion.

3) The link cost is mostly determined by the fiber cable and the cost of its installation (particularly when the cable is directly buried into the ground), while repeater costs only play a minor role. As a consequence, the cost c of a channel-kilometer can be roughly given by

$$c \approx c_0/N^a \quad , \qquad (6.3)$$

where c_0 is the fixed cost, N is the number of transmitted channels, and a is a parameter close to unity. At least in the initial phase of its development, the cost of the fiber system hence differs from its conventional counterpart whose economy of scale is less spectacular ($c \sim N^{-\frac{1}{2}}$) [6.6].

Let us now turn to the optical fiber link, and observe that its performance is basically limited by two factors: the signal-to-noise ratio and intersymbol interference, or equivalently attenuation and dispersion. The attenuation within the system dominates at low pulse repetition rates, when the bits are so widely separated that the signal-to-noise ratio drops to unacceptable levels before dispersion can play a part. On the other hand, at high repetition rates, dispersion produces pulse "tails" that interfere with neighbouring pulses. Thus, for example, a transmitted "0" may be identified as a "1", or vice versa. We will consider these two mechanisms separately.

Firstly we assume the absence of inter-symbol interference, so that we can simply write down the equation for the inter-repeater distance

$$\alpha L + K a_j = 10 \lg(P_t/P_r) \quad , \tag{6.4}$$

where α is the attenuation of the fiber [dB/km], a_j is the average splice loss [dB], K is the number of splices (= L/L_0), P_t is the transmitted power (effective), and P_r is the minimum received power for which acceptable receiver performance is achieved. The quantity L_0 could be interpreted to be the manufacturing length, but it would probably be better if L_0 would represent the average distance between splices at the end of the cable or system life. This definition is particularly reasonable if we bear in mind that it is the fate of telecommunications cables to be damaged during their operational life (by mechanical diggers, etc.).

The quantity of interest in (6.4) is the repeater spacing L. A convenient way to determine L is to define the dimensionless length $\ell = L/L_0$, and rewrite (6.4) as follows

$$\ell = \frac{10}{\alpha L_0} \lg \frac{P_t}{P_r} - \frac{a_j}{\alpha L_0} [\ell] \ . \tag{6.5}$$

where $[\ell]$ represents the integer part of ℓ.

Example 6.1. Let $10 \lg(P_t/P_r) = 50$ dB, $\alpha = 5$ dB/km, $L_0 = 1$ km, $a_j = 0.5$ dB (very pessimistic). Solving (6.5) iteratively, we obtain L = 9.1 km, so that the total splice loss adds up to 4.5 dB, while the fiber loss is responsible for 45.5 dB. This example shows that splice losses can be significant and must be considered.

Dispersion becomes decisively significant when the delay distortion of the pulse becomes as large as the bit interval. It is possible to compensate for the effects of dispersion, but only at the expense of enhanced noise. This procedure, known as equalization, forces the pulses to return to zero more rapidly by the use of specially tailored high-pass or "differentiating" circuits. It follows that the accentuated high frequency response will also boost high frequency noise. Eventually, beyond a certain limit, any increase in the bit rate causes a sharp decrease in the repeater spacing. This dispersion limit can be estimated from the equation

$$\sigma_{tot}L = \frac{1}{4f_0} , \tag{6.6}$$

where σ_{tot} is the root-mean-square value of the delay distortion per unit length, and f_0 is the modulation rate. It has been assumed that the dispersion increases linearly with the fiber length L, and that the dispersion limit corresponds to a performance impairment of 1 dB.

Mode mixing, which reduces dispersion to some extent, can be modelled by replacing the actual length by an effective length L_{ef}

$$\begin{aligned} L_{ef} &= \sqrt{L_c L} , \quad L > L_c \\ &= L , \quad L < L_c , \end{aligned} \tag{6.7}$$

where L_c is the critical length within which a stable mode distribution is reached. Typically, L_c may range from hundreds of meters to several kilometers, and usually has to be determined experimentally. A more general model would be

$$L_{ef} = L_c^{1-p} L^p , \tag{6.8}$$

where the exponent p can be found by curve fitting to measured impulse responses. The situation described by (6.6) basically accounts for modal dispersion which is dominant when a narrow spectral source, such as a laser diode, is used. However, in many applications a light-emitting diode is sufficient. The broad spectral width of such diodes gives rise to chromatic or material dispersion, so that $\sigma_{tot}L$ in (6.6) must be written in the form

$$\sigma_{tot}^2 L^2 = \sigma_{mod}^2 L_{ef}^2 + \sigma_{chr}^2 L^2 , \tag{6.9}$$

where σ_{mod} and σ_{chr} are the root-mean-square intermodal and chromatic delay distortions, respectively. The latter may be measured separately or estimated on the basis of the Sellmeier equation [6.7].

If we now use (6.4) and (6.6), we can estimate the modulation rate limit, f_{cr}, beyond which the fiber link is dispersion limited

$$f_0 > f_{cr} = \frac{\alpha}{4\sigma_{tot}} \frac{1}{10\ \lg(P_t/P_r)} \quad . \tag{6.10}$$

Example 6.2. Using a laser diode source (negligible chromatic or material dispersion), a graded index fiber with $\sigma_{tot} = 2$ ns/km, $\alpha = 11$ dB/km, and a receiver for which $10\ \lg(P_t/P_r) = 50$ dB, a dispersion limit of 27.5 MBd is obtained. On the other hand, with a LED source and assuming $\sigma_{tot} = 4$ ns/km, $\alpha = 5$ dB/km, the dispersion limit is reduced to 6.25 MBd.

For modulation rates below the critical value, we have power limitation and (6.4) is applicable, while above this value the repeater distance can be estimated from (6.6). We would like to emphasize that the formulae given here are rather coarse and qualitative in nature. However, they are useful as mental models for systems design, and are reasonably accurate for experimentally determined parameters. A more accurate analysis will be discussed later.

To clarify the use of the formulae we have just discussed, let us consider a link which is to have the following properties:

Bit rate: $R = 2.048$ Mbit/s,
Line code: 3B4B, $f_0 = 2.731$ MBd,
Waveguide: graded index, $\alpha = 5.0$ dB/km,
Numerical aperture: 0.18,
Manufacturing length: $L_0 = 1.0$ km,
Source: LED, $\lambda = 900$ nm, $P_1 = 3.91$ dBm[2].

The first question we must answer is whether the link will be attenuation or dispersion limited. To do this we must know P_r, the minimum required power for acceptable performance. Consequently, we must know the sensitivity of the receiver, which in turn strongly depends on the receiver structure. This

[2] In systems design, powers are typically expressed as a certain number of dB referred to some specified level. When the reference level is 1 mW, the "power level" is expressed in dBm (m for mW).

complex question will be considered a little later, and for now we will simply assume that, somehow, we know the receiver sensitivity to be, say, -71.42 dBm[3]. The use of (6.10) then shows us that the link is power limited. Consequently, we go to (6.4) or (6.5) and solve iteratively for the repeater spacing L. A convenient means of illustrating the effect of the various factors is to use the concept of the power budget. The power budget for our example is shown in Table 6.2, from which we see that, in addition to our previous specifications, we have assumed a source coupling loss of 18.13 dB, a splice loss of 0.3 dB/splice, a detector coupling loss of 0.5 dB, and an impairment allowance of 5 dB.

Table 6.2. The power budget for the example in the text

Transmitter:		
Light-emitting diode (λ = 0.9 μm)	P_1 =	3.91 dBm
Coupling loss into waveguide	k_1 =	18.13 dB
Waveguide input level	$P_1' = P_1 - k_1$ =	-14.22 dBm
Receiver:		
Avalanche photodiode, sensitivity	P_2 =	-71.42 dBm
Coupling loss from waveguide	k_2 =	0.5 dB
Waveguide output level	$P_2' = P_2 + k_2$ =	-70.92 dBm
Transmission loss:	$P_1 - P_2'$ =	56.70 dB
Waveguide loss, α = 5.0 dB/km, 9.8 km	L =	49.0 dB
Splice losses, a_j = 0.3 dB (9 splices)	a_jK =	2.7 dB
Impairment allowance	k_r =	5.0 dB
	$\alpha L + a_jK + k_r$ =	56.7 dB
Repeater distance	L =	9.8 km

Let us now discuss some line codes suitable for optical fiber links. As already mentioned, we require frequent signal transitions to ease clock extraction, and at the same time we must restrict our signal alphabet to zero

[3] See footnote 2 on page 189.

and positive values. In addition, the line code should be such as to allow link supervision, the location of faulty repeaters, and exploitation of the alarm and control functions of the interface code (HDB-3 or CMI). In the case of dispersion limited links, we also require that the code should not significantly increase the transmission rate.

Table 6.3. The simplest means of optically coding HDB-3 or AMI signals

AMI/HDB-3	Optical power
-1	0
0	P_t
+1	$2\ P_t$

The simplest code that fulfills these requirements is shown in Table 6.3. In this code, a power level P_t represents zero, while levels $2P_t$ and 0 represent +1 and -1, respectively. In other words, the coding operation simply requires the source output to be biased to a level P_t, followed by modulation by the HBD-3 signals, as illustrated in Fig.6.2. The main disadvantage of this method is the necessity for level stabilization at the transmitter and receiver. This is sufficiently serious to force us to look at other alternatives.

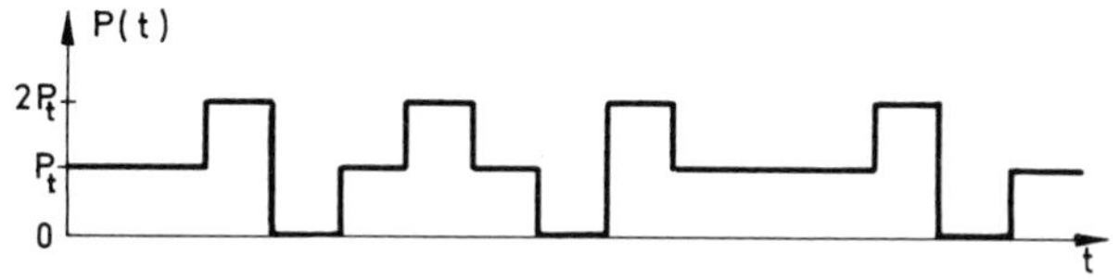

Fig. 6.2. A typical waveform for the line code of Table 6.3: interface code HDB-3, and bits to be transmitted 00+-0+0-+000+-0

At the next level of complexity, we have the class known as 1B2B (one bit represented by two bits). Examples of such codes are the 2-level AMI[4] (Alternative Mark Inversion), the CMI (Complemented Mark Inversion), and the bi-

[4] The AMI code (sometimes used in conventional electrical systems) is identical to HDB-3 except that "violation" bits are not used. Thus, long sequences of zeroes or ones are allowed to occur.

phase. The characteristics of these codes are shown in Table 6.4. We see that using the zero-one optical alphabet available to us, each electrical digit (+1, 0, -1) is re-coded into two optical bits. Thus, in the two-level AMI code, a +1 is coded as 11, a -1 as 00, and zeroes after -1 as 10, while zeroes after +1 as 01. (Note that the optical ones and zeroes correspond to "light" and "no light" respectively). The waveforms for this and the other codes in Table 6.4 are shown in Fig.6.3.

Table 6.4. Characteristics of some IB2B codes

Data	AMI/HDB-3	2-level AMI	CMI	Bi-phase
1	+	1 1	1 1	1 0
	-	0 0	0 0	
0	0	1 0 after -	1 0	0 1
		0 1 after +		

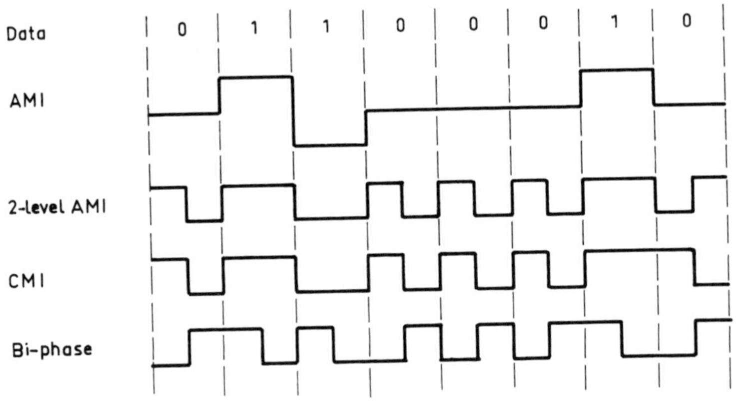

Fig. 6.3. Typical waveforms for the line codes of Table 6.4, and the AMI interface code

All these codes are inherently well-balanced (implying no long runs of zeroes or ones) as can be seen from their digital sums: ± 2 for 2-level AMI and bi-phase, and ± 3 for CMI. Moreover, they allow pair synchronization at the receiver by identification of the following forbidden combinations: 01 after -, 10 after + for 2-level AMI, 01 for CMI, and 11 or 00 for bi-phase.

The main disadvantage of these codes (indeed all 1B2B codes) is that they double the modulation rate and are therefore only suitable for attenuation limited systems.

For dispersion limited links, any increase in the modulation rate is undesirable, so that 1B2B codes are not normally acceptable. Although codes of the type 2B3B, 3B4B, 5B6B etc. have been studied and implemented [6.8], and they do not increase the transmission rate as much as the 1B2B, the most efficient method involves codes that do not increase the rate at all. These fall into the class 1B1B and include such techniques as data scrambling using self-synchronizing feed-back shift registers [6.9], and delay modulation [6.10]. Another alternative is to use the so-called Q-level codes ($Q > 2$) [6.11], which reduce the transmission rate by $\lg_2 Q$ (at most).

Our next problem is to evaluate the *receiver sensitivity*, which is a system parameter of fundamental importance, as given in (6.4) and (6.10). In order to do this, let us first model the system as in Fig.6.4. Pulses, whose shape is given by $f_T(t)$ and spectrum by $F_T(j\omega)$, are transmitted into a fiber, whose power impulse response is assumed to be $h_c(t)$, corresponding to a transfer function $H_c(j\omega)$. Assuming power linearity in the waveguide, the detected signal pulse then takes the form:

$$P_s(t) \simeq f_T(t) * h_c(t) \quad ,$$
$$= \int_{-\infty}^{\infty} f_T(t')h_c(t - t')dt' \quad . \tag{6.11}$$

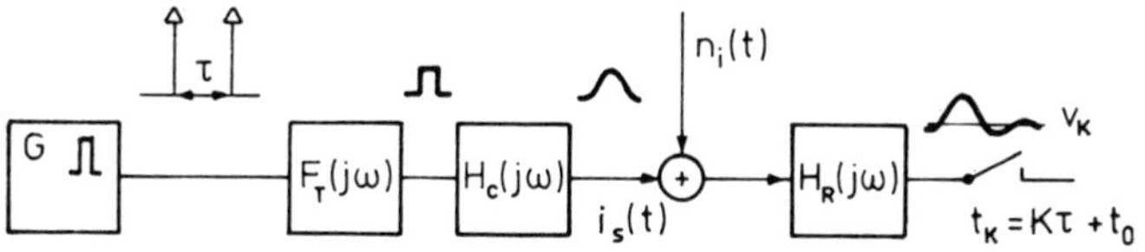

Fig. 6.4. A módel of the optical fiber system

In reception, the photodetector converts this optical power into a corresponding current $i_s(t)$, which is corrupted by the equivalent noise current of the receiver. This noise current can be evaluated from the equivalent circuit of the receiver input (Fig.6.5). The way in which the average signal current depends on the optical signal power was discussed in Sect.2.3 and is reproduced here for convenience

$$i_s(t) = \frac{M\eta e\lambda}{hc} \cdot P_s(t) \quad , \tag{6.12}$$

where M is the multiplication factor of the detector, η its quantum efficiency, e the electronic charge, λ the wavelength of the radiation, h Planck's constant, c the velocity of light, and $P_s(t)$ the signal power. The current in (6.12) is filtered by the input network so that the Laplace transform of the signal voltage at the preamplifier input becomes

$$U_i(s) = \frac{I_s(s)R_b}{1 + s\tau_i} \quad , \tag{6.13}$$

where $\tau_i = R_bC_i$ is the time constant of the input circuit, and R_b is its equivalent parallel resistance. If the current $i_s(t) = i_{s0}$ would be constant, the equivalent noise current generator can be shown to have a spectral power density (when using a field effect transistor) given by

$$S_{ni}(\omega) = MeFi_{s0} + \frac{2kT}{R_b} + \frac{2kT(0.7)}{g_mR_b^2}(1 + \omega^2\tau_i^2) \quad , \tag{6.14a}$$

where, [cf. (4.31)]

$$F = k'M + (2 - 1/M)(1 - k') \tag{6.14b}$$

is the avalanche photodetector noise factor (Sect.4.4), k is Boltzmann's constant, and g_m is the mutual conductance of the field-effect transistor.

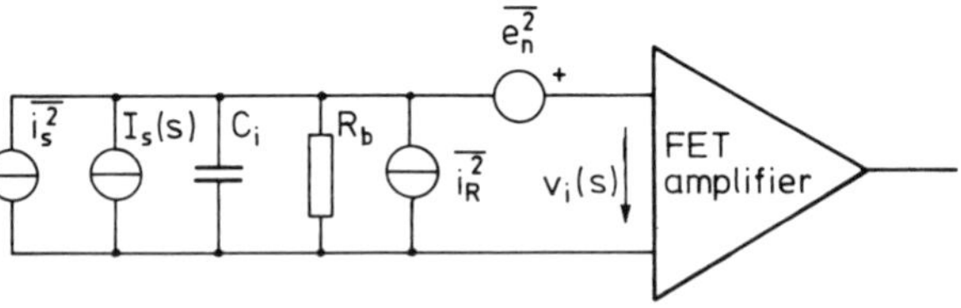

Fig. 6.5. Equivalent circuit of the receiver front-end for a FET input stage

In passing, we would like to note that, even though the noise properties of a bi-polar transistor are more complex, they only affect the last term in (6.14). Based on the simplified noise models of VAN DER ZIEL [6.12], this term takes the following approximate form

$$S_{n,bip.} = \frac{2kT}{R_b^2}\,[r_b + r_e/(2\alpha_F)]\,\frac{(1+\omega^2\tau_i^2)(1+\omega^2\tau_b^2)}{(1+\omega^2\tau_e^2)} \quad , \tag{6.15}$$

where r_b and r_e are the base and emitter resistances of the transistor model, while α_F is the collector-to-emitter current ratio. The terms τ_b and τ_e are given by

$$\left.\begin{aligned} \tau_b &= \frac{r_b c_i}{1-\alpha_F} \approx \beta r_b c_i \quad , \\ \tau_e &= \frac{r_e c_i}{1-\alpha_F} \approx \beta r_e c_i \quad , \end{aligned}\right\} \tag{6.16}$$

where $\beta = \alpha_F/(1-\alpha_F) \gg 1$. It should be observed that, since $r_e = kT/eI_E$, the preamplifier noise contribution can be minimized by selecting the proper value for I_E.

Returning now to the main theme, we observe that, as in Sect.4.4, what we are interested in is the signal-to-noise ratio. The mean value of the received and equalized signal sample is simply given by

$$\bar{v}_\ell = \int_{-\infty}^{\infty} h_R(t_\ell - t')i_s(t')dt' \quad , \tag{6.17}$$

where $h_R(t)$ is the impulse response of the receiver and equalizer, and the index ℓ refers to the sampling moment. In contrast, evaluation of the sample variance involves a few more steps, and can be obtained as follows.

First, we observe that the correlation function, corresponding to $S_{ni}(\omega)$ in (6.14), can be obtained by taking the inverse Fourier transform of $S_{ni}(\omega)$. This operation gives the following result

$$K_{ni}(\tau) = \left[MeFi_{s0} + \frac{2kT}{R_b}\left(1 + \frac{0.7}{g_m R_b}\right)\right]\delta(\tau) - \frac{2kT(0.7)}{g_m R_b^2}\,\tau_i^2\ddot{\delta}(\tau) \quad , \tag{6.18}$$

where $\delta(\tau)$ is the Dirac delta function, and $\ddot{\delta}(\tau)$ is its second derivative. Both of these should be understood in the sense of the mathematical theory of distributions.

With the aid of (6.18), we can now define the following time-varying correlation function for the input noise

$$K_{ni}(t,\tau) = \overline{n_i(t)n_i(t+\tau)}$$

$$= \left[MeFi_s(t) + \frac{2kT}{R_b}\left(1 + \frac{0.7}{g_mR_b}\right)\right]\delta(\tau) - \frac{2kT(0.7)\tau_i^2}{g_mR_b^2} \cdot \ddot{\delta}(\tau) \quad , \tag{6.19}$$

where the bar denotes the expected value.

The effect of the receiver "filter" is to convolve the input noise $n_i(t)$ with the receiver impulse response $h_R(t)$, so that the output noise variance, at the sampling moment $t_\ell = \ell\tau + t_0$, becomes

$$\sigma_\ell^2 = \overline{n_0^2(t)} = \int_{-\infty}^{\infty}\int h_R(t-t')h_R(t-t'')\overline{n_i(t')n_i(t'')}dt'dt'' \quad ,$$

$$= \int_{-\infty}^{\infty}\int h_R(t-t')h_R(t-t'-t''')K_{ni}(t',t''')dt'dt''' \quad , \tag{6.20}$$

where $t'' = t' + t'''$. Use of the expression for $K_{ni}(t,\tau)$ in (6.19) eventually leads to the following result

$$\sigma_\ell^2 = MeF\int_{-\infty}^{\infty} dt' h_R^2(t_\ell - t')i_s(t') + \frac{2kT}{R_B}\left(1 + \frac{0.7}{g_mR_b}\right)\int_{-\infty}^{\infty} dt\, h_R^2(t)$$

$$+ \frac{2kT(0.7)\tau_i^2}{g_mR_b^2}\int_{-\infty}^{\infty} \dot{h}_R^2(t)dt \quad , \tag{6.21}$$

where $h_R(t) = dh_R(t)dt$. Here, it should be noted that this variance is equivalent to the one described by PERSONICK [6.13] and GOELL [6.14] and is signal dependent. Indeed, at the desired low error rate and with proper receiver design, the signal-dependent noise becomes the largest component.

The signal-to-noise ratio κ can now be evaluated by using (6.17) and (6.21), which yield [cf.(4.12)]

$$\kappa = \frac{\bar{v}_\ell^2}{\sigma_\ell^2} = \frac{Q_s^2B_1^2}{MeFQ_sB_1'^2 + \frac{kT}{R_b}\left(1 + \frac{0.7}{g_mR_b}\right)B_2 + \frac{kT(0.7)}{g_mR_b^2}\tau_i^2\frac{4\pi^2}{3}B_3^3} \quad , \tag{6.22}$$

where Q_s represents the charge transported by a single current pulse $i_s(t)$.

The bandwidths B_1, B_1', B_2, B_3 in (6.22) are given by

$$B_1 = \frac{\int_{-\infty}^{\infty} h_R(-t')i_s(t')dt'}{2H_R(0)Q_s} = \frac{\int_{-\infty}^{\infty} H_R(j2\pi f)I_2(j2\pi f)df}{2H_R(0)Q_s} , \tag{6.23}$$

$$B_1'^2 = \frac{\int_{-\infty}^{\infty} h_R^2(-t')i_s(t')dt'}{4H_R^2(0)Q_s} , \tag{6.24}$$

$$B_2 = \frac{\int_{-\infty}^{\infty} h_R^2(t)dt}{2H_R^2(0)} = \frac{\int_{-\infty}^{\infty} |H_R(j2\pi f)|^2 df}{2H_R^2(0)} , \tag{6.25}$$

$$B_3^3 = \frac{3}{8\pi^2 H_R^2(0)} \int_{-\infty}^{\infty} \dot{h}_R^2(t)dt = \frac{3}{H_R^2(0)} \int_0^{\infty} f^2 |H_R(j2\pi f)|^2 df . \tag{6.26}$$

These bandwidths have been defined such that, if $i_s(t) = Q_s\delta(t)$, and $h_R(t)$ is the impulse response of an ideal filter of bandwidth W, all bandwidths become equal to W.

What we are looking for is the signal power necessary to achieve a certain signal-to-noise ratio, or equivalently to ensure a certain bit error rate. Thus, in (6.22), we assume knowledge of κ and solve the resultant quadratic equation in Q_s. If we further assume that the bit interval has a width $\tau = 1/f_0$, then the average current is Q_s/τ, and it can be related back to the optical power via (6.12). Using this approach, it can be shown that the optical power corresponding to a given signal-to-noise ratio takes the form:

$$P_s = \frac{E_s}{\tau} = \frac{P_z}{2} \left(1 + \sqrt{1 + 4P_{cr}/P_z}\right) , \tag{6.27a}$$

where

$$P_z = \kappa F \frac{hc}{\eta\tau\lambda} \frac{B_1'^2}{B_1^2} . \tag{6.27b}$$

Here, P_z represents an equivalent optical power such that P_s/P_z is equal to $(1/\kappa)$ times the final ratio of the electrical signal power to the signal-dependent noise power. The quantity

$$P_{cr} = \frac{kTB_2}{eFR_b B_1'^2 M^2} \left[1 + \frac{0.7}{g_m R_b} \left(1 + \frac{4\pi^2}{3} \frac{\tau_i^2 B_3^3}{B_2}\right)\right] \frac{hc}{e\eta\lambda\tau} , \tag{6.28}$$

describes that signal power for which the signal-dependent noise equals the joint contributions of the pre-amplifier thermal and shot noises.

Example 6.3. Assume R = 34,4 Mbit/s, W = R/2 (ideal low pass filter), λ = 0.85 μm, R_b = 1MΩ, C = 5 pF, M = 50. Then P_s = 2.3 nW (-56.3 dBm), P_z = 2.2 nW, and P_{cr} = 134 pW, for $\kappa \simeq 21.1$ dB, corresponding to an error rate of 10^{-9}. We see that the formalism which has been used allows easy identification of the different noise sources.

We now have the basic tools necessary for the evaluation of receiver sensitivity. Our next task is to optimize receiver performance by the optimum selection of the decision threshold, the receiver front-end, and the equalizer. In addition, the influence of inter-symbol interference must also be eliminated.

Initially, we will assume inter-symbol interference to be negligible. We will also assume that the received sample v obeys normal (Gaussian) statistics and that its mean and variance are different for the two transmitted states 1 (power on) and 0 (power off)

$$\left.\begin{aligned} 1:\quad \bar{v} &= V \;, & \operatorname{var} v &= \sigma_1^2 \;, \\ 0:\quad \bar{v} &= 0 \;, & \operatorname{var} v &= \sigma_0^2 \;. \end{aligned}\right\} \tag{6.29}$$

In other words, when a "1" has been transmitted, the mean of the received sample is V and its variance σ_1^2, while for a transmitted zero, the mean is 0 and the variance σ_0^2. In fact, the assumption of normality is not quite correct because of two factors:

1) Discrete electron-hole pair generation, which takes place in the detector, is most accurately modelled as an inhomogeneous Poisson process, which is subsequently filtered by the reactive (energy storage) elements in the circuitry.

2) When avalanche multiplication occurs, the Poisson distribution is further modified such that the variance becomes greater than expected.

However, in our application, because thousands of photons are needed for reliable detection, the Gaussian assumption becomes reasonably good, particularly if the decision threshold is also evaluated on the same basis. (The actual threshold would, of course, have to be experimentally adjusted during receiver tuning). The advantage we gain is that the Gaussian assumption enables us to include the thermal and shot noises in the same probability density. The probability density (Fig.6.6) at the input of the decision circuit can then be written in the form [6.15]

$$\left.\begin{aligned} p(v/1) &= \frac{1}{\sqrt{2\pi}\sigma_1} \exp\left[-(v - V)^2/2\sigma_1^2\right] \quad , \\ p(v/0) &= \frac{1}{\sqrt{2\pi}\sigma_0} \exp\left[- v^2/2\sigma_0^2\right] \quad . \end{aligned}\right\} \tag{6.30}$$

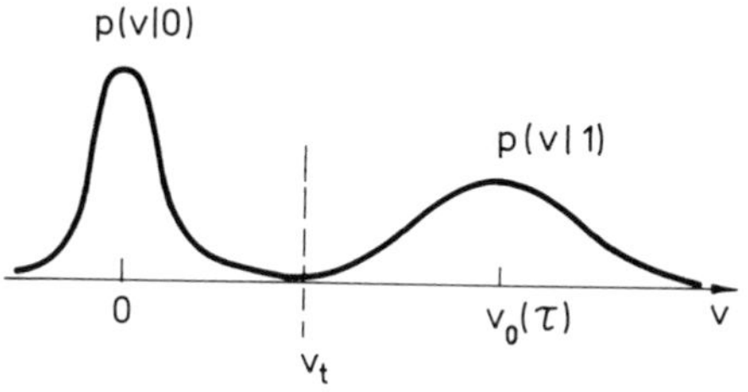

Fig. 6.6. Conditional probability density at the input of the decision circuit of a binary receiver

The mean value in the above is given by (6.17) while the variances are given by (6.21). If, in addition, we assume that for σ_0 the term containing i_s is zero, then the use of (6.17) and (6.21) gives us

$$\begin{aligned} V &= B_1 Q_s \quad , \\ &= MB_1 \cdot \frac{\eta e \lambda \tau}{hc} \cdot P_s \quad , \end{aligned} \tag{6.31}$$

$$\left.\begin{aligned} \sigma_0^2 &= eFM^2 B_1'^2 \cdot \frac{e\eta\lambda\tau}{hc} \cdot P_{cr} \quad , \qquad \text{and} \\ \sigma_1^2 &= eFM^2 B_1'^2 \cdot \frac{e\eta\lambda\tau}{hc} (P_s + P_{cr}) = \sigma_0^2 + \sigma_s^2 \quad . \end{aligned}\right\} \tag{6.32}$$

Here σ_s^2 represents the signal-dependent variance [$MeFQ_s B_1'^2$ according to (6.22)].

The average error probability can now be minimized using the following decision rule [6.16]

$$\Lambda(v) = \frac{p(v/1)}{p(v/0)} \underset{0}{\overset{1}{\gtrless}} \frac{p_0}{p_1} \tag{6.33}$$

where $\Lambda(v)$ is the likelihood ratio and p_0 and p_1 are the a priori probabilities of the states 0 and 1. (For balanced codes $p_0 = p_1 = 0.5$). Taking logarithms, an equivalent decision rule is obtained

$$\ell(v) = \ln \Lambda(v) \quad ,$$

$$= \ln \frac{\sigma_0}{\sigma_1} + \frac{v^2}{2\sigma_0^2} - \frac{(v - V)^2}{2\sigma_1^2} \mathop{\gtrless}_{0}^{1} \ln \frac{p_0}{p_1} \quad , \tag{6.34}$$

where the inequality is interpreted such that a 1 is assumed to be transmitted, if (left side) > (right side) and a 0 if (left side) < (right side). For equality either can be chosen. The decision threshold thus corresponds to the equality, and can be evaluated to be:

$$v_t = V \left[- \frac{\sigma_0^2}{\sigma_1^2 - \sigma_0^2} \pm \frac{\sigma_0 \sigma_1}{\sigma_1^2 - \sigma_0^2} \sqrt{1 + \frac{\sigma_1^2 - \sigma_0^2}{V^2} \ln \left(\frac{\sigma_1^2}{\sigma_0^2} \frac{p_0^2}{p_1^2} \right)} \right] \tag{6.35}$$

The negative root is not useful and will be neglected. Setting $\sigma_1^2 = \sigma_s^2 + \sigma_0^2$, and $p_0 = p_1$ we obtain

$$v_t = \frac{V}{\sigma_s^2} \sigma_0^2 \left[-1 + \frac{\sigma_1}{\sigma_0} \sqrt{1 + \frac{\sigma_s^2}{V^2} \ln \left(1 + \frac{\sigma_s^2}{\sigma_0^2} \right)} \right] \quad . \tag{6.36}$$

In terms of P_s and P_{cr} [from (6.31) and (6.32)] we thus have the following expression for the threshold

$$\frac{v_t}{V} = \frac{P_{cr}}{P_s} \left[-1 + \sqrt{\frac{P_s + P_{cr}}{P_{cr}}} \sqrt{1 + \frac{P_z}{\kappa P_s} \ln \left(\frac{P_s + P_{cr}}{P_{cr}} \right)} \right] \quad ,$$

$$\approx \frac{\sqrt{P_{cr}} \left(\sqrt{P_{cr} + P_s} - \sqrt{P_{cr}} \right)}{P_s} \quad . \tag{6.37}$$

The error probability to be expected using this threshold can be evaluated on the basis of probability densities from the relationship [6.15]

$$P[\varepsilon] = p_1 Q \left(\frac{V - v_t}{\sigma_1} \right) + p_0 Q \left(\frac{v_t}{\sigma_0} \right) \quad . \tag{6.38}$$

Again setting $p_0 = p_1 = 1/2$, we obtain the following approximate formula

$$P[\varepsilon] \approx Q \left[\frac{V(\sigma_1 - \sigma_0)}{\sigma_s^2} \right] \quad . \tag{6.39}$$

In the above, Q(.) is the complementary probability integral given by:

$$Q(x) = \int_x^\infty dt \frac{e^{-t^2/2}}{\sqrt{2\pi}} \lessapprox \frac{e^{-x^2/2}}{\sqrt{2\pi}\, x} \quad . \tag{6.40}$$

with the latter bound good for large x [6.17]. The inverse function can be evaluated by setting $Q[x_q] = q$, with the result

$$x_q^2 \approx y_0 - \frac{\ln y_0}{1 + 1/y_0} \quad , \quad y_0 = -\ln(2\pi q^2) \quad . \tag{6.41}$$

The signal power can now be found by solving (6.38) or (6.39) for P_s, or equivalently for Q_s. On the basis of (6.39), and using (6.27,31,32 and 37) we obtain

$$x_q^2 = \frac{\kappa}{4} = \left[\frac{V(\sigma_1 - \sigma_0)}{\sigma_s^2}\right]^2 = \frac{P_s\kappa}{P_z} \frac{\left(\sqrt{P_s + P_{cr}} - \sqrt{P_{cr}}\right)^2}{P_s} \quad , \tag{6.42}$$

leading to the following expression for signal power

$$P_s = \frac{P_z}{4} + \sqrt{P_z P_{cr}} \quad . \tag{6.43}$$

This equation predicts a result that is typically a factor of 2 better than the one given by (6.27). Using the numerical values of Example 6.3, we find that the necessary power is now reduced from 2.3 nW to 1.1 nW, or by 3.2 dB. Moreover, the optimum threshold level is at 0.25 V instead of the 0.5 V of the conventional case ($\sigma_0 = \sigma_1$). If $\sigma_0 = 0$, the optimum threshold is at zero and $P_s = P_z/4$, so that the improvement with respect to (6.27) is 6 dB.

Our next optimization is related to the choice of the multiplication factor M, which varies according to the bias voltage. The objective is to maximize the signal-to-noise ratio κ in (6.22). Since the charge Q is related to the optical energy $E_s = P_s\tau$ by a constant factor $M\eta e\lambda/hc$ [see (6.12)], the numerator and denominator in (6.22) must be proportional to M^2 and M^3, respectively. Thus, an optimum multiplication factor, say M_{opt}, must exist and can be found by taking the derivative of κ with respect to M, and equating it to zero. Then, using (6.28) and defining $P_0 = M^2FP_{cr}$, the following condition is obtained

$$M^3 + (1 - k')M/k' - 2P_0/(k'P_s) = 0 \quad . \tag{6.44}$$

This equation can be solved using Cardano's formulae [6.17], and we obtain

$$\left.\begin{aligned} M_{opt} &= \left(\sqrt{q^3 + r^2} - r\right)^{1/3} - \left(\sqrt{q^3 + r^2} + r\right)^{1/3} \quad , \quad \text{where} \\ q &= \frac{1 - k'}{3k'} \quad , \quad \text{and} \quad r = -\frac{P_0}{k'P_s} \quad . \end{aligned}\right\} \tag{6.45}$$

In this formula, we set M = 1, if M_{opt} becomes less than unity. The condition for this can be shown to be

$$P_s > 2P_0 \quad . \tag{6.46}$$

For k' = 1 (e.g., in a germanium detector), we find that the signal-dependent noise in (6.21) and (6.22) is twice the value of the thermal plus shot noise contributions. For $k' \ll 1$, the signal-dependent noise is even larger.

Our next optimization is related to the choice of the resistance R_b in the bias circuit (Fig.6.5). For maximum sensitivity, R_b should tend to infinity, as can be seen from (6.22). This being unrealistic, an acceptable compromise is to choose R_b such as to make the contributions of the thermal and amplifier shot noises equal. Thus, equating the last two terms in the denominator of (6.22), we obtain

$$R_B > R_{b,cr} = \frac{3}{4\pi^2(0{,}7)} \frac{g_m}{C_i^2} \frac{B_2}{B_3^3} \quad . \tag{6.47}$$

This result simplifies the expression for P_{cr} in (6.28), so that P_0 becomes

$$P_{0,min} = P_{cr}FM^2 = \frac{8\pi^2(0{,}7)}{3} \cdot \frac{kT}{e} \cdot \frac{C_i^2}{g_m} \cdot \frac{B_3^3}{B_1'^2} \cdot \frac{hc}{e\eta\lambda\tau} \quad . \tag{6.48}$$

We see that P_0 can be minimized by tailoring the detector and preamplifier circuits to have maximum g_m/C_i^2. Hence, g_m/C_i^2 can be taken as a figure of merit of the receiver.

Example 6.4. Let us assume g_m = 10 mS, $f_0 = \tau^{-1}$ = 34,368 MBd, $B_1 = B_1' = B_2 = B_3 = f_m/2$, η = 0,8, λ = 850 nm, C_i = 10 pF, $P(\varepsilon) = 10^{-9}$, and k' = 0,02. Solving for $R_{b,cr}$ in (6.47) gives R_b = 37 kΩ. Iteration of (6.43) and (6.44) leads to the simultaneous solutions M_{opt} = 47 and P_s = 4.6 nW (or $E_s = 134 \times 10^{-18}$J). With these values we obtain P_z = 8.3 nW and P_{cr} = 0.8 nW. To estimate the required P_s at a different modulation rate, we invoke the relationship $P_s = E_s f_0$ and write it in the form

$$10\ \lg\left(\frac{P_s}{\mathrm{mW}}\right) = 10\ \lg\left(\frac{E_s}{\mathrm{mJ}}\right) + 10\ \lg\left(\frac{f_0}{\mathrm{Bd}}\right) \quad . \tag{6.49}$$

For our example, $E_s = 134 \times 10^{-18}$J, so that the first term on the right-hand side of (6.49) becomes -128.7. In fact, this relationship is oversimplified because P_z, M_{opt}, and P_0 all depend on the bandwidth.

Except for the various bandwidths in (6.23) to (6.26), our analysis of the attenuation-limited receiver is now almost complete. These bandwidths are determined by the actual circuit configuration that is used and must therefore be separately evaluated for any given receiver. However, before giving an example of a typical realizable filter, let us determine our ultimate performance limit. This occurs for a receiver with perfect conversion efficiency and noise-free characteristics but which produces errors because of the quantum nature of photons. The probability of error for a transmitted "1" is then given by [6.18]

$$P[\varepsilon/1] = \exp(-E_s/h\nu) \quad . \tag{6.50}$$

In our ideal case, we can neglect the probability of error for zero transmission because of our noise-free receiver. Equation (6.50) is hence sufficient for the estimation of the energy necessary for a given error rate. Selecting λ = 850 nm and stipulating an error probability of 10^{-9}, we conclude that at least 20 photons, corresponding to an energy level of -140,0 dB(mJ), are required. Comparing this result to the energy required in Example 6.4, we see that the ideal receiver is better by about 11.3 dB. These results are summarized in Fig.6.7 for two values of C_i. The modulation rate dependence has been evaluated using (6.49).

As an example of a simple realizable filter, we can look at the RC network formed by the detector circuit itself. In order to keep the noise bandwidth B_3 in (6.26) finite, we must include another time constant τ_2, which be realistically assumed to exist somewhere within the preamplifier circuit. Thus, let us suppose that the overall transfer function is given by

$$H_R(j\omega) = \frac{R_b}{(1 + j\omega\tau_i)(1 + j\omega\tau_2)} \quad , \tag{6.51}$$

where $\tau_i = R_b C_i$ as in (6.13). For an input current step of height I_0, the output signal becomes

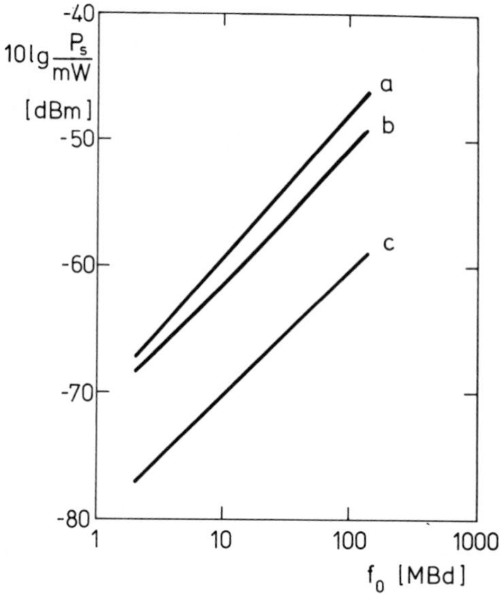

Fig. 6.7. Receiver sensitivity as a function of modulation rate for the parameters of Example 6.4: (a) C_i = 10 pF, (b) C_i = 1 pF, and (c) ideal quantum receiver

$$v_0(t) = \frac{I_0R_b}{\tau_i - \tau_2}\{\tau_i[1 - \exp(-t/\tau_i)] - \tau_2[1 - \exp(-t/\tau_2)]\} \quad ,$$

$$\approx \frac{I_bR_b(t - \tau_2)}{\tau_i - \tau_2} \quad \text{for} \quad \tau_2 << t << \tau_i \quad . \qquad (6.52)$$

It can be seen that the output is an integrated version of the input, provided that the integration time satisfies the condition in (6.52). In the case of a pulse of width τ_s, the corresponding condition is simply obtained by replacing t by τ_s. Use of this new condition in (6.23) through (6.26) then yields the following results for the bandwidths in (6.22)

$$\begin{aligned} B_1 &\approx (1 - \tau_2/\tau_s)/(2\tau_i) \quad , \\ B_1' &\approx 1/\tau_i \quad , \\ B_2 &\approx 1/4\tau_i \quad , \\ B_3^3 &\approx 3/(16\pi^2\tau_i^2\tau_2) \quad . \end{aligned} \qquad (6.53)$$

To complete the filter design, we place a constraint on τ_2, and require that it satisfy the condition

$$0.7\ C_i/g_m < \tau_2 < \tau_s/3 \quad . \qquad (6.54)$$

The upper bound in (6.54) has been obtained by first evaluating P_s in (6.43) by using (6.53), and then optimizing P_s with respect to τ_2/τ_s. The lower bound on τ_2 has been estimated by inserting (6.53) into (6.47), and selecting that value of R_b for which the thermal noise equals the pre-amplifier shot noise. As a result, the lower bound limits the shot-noise contribution from the pre-amplifier, while the upper bound prevents excessive distortion of the integrated pulse.

Example 6.5. A typical receiver of this type may have C_i = 10 pF, R_b = 1MΩ, g_m = 10 mS, and τ_s = 20 ns. Then, the optimum value of τ_2 would be 1 ns, and the receiver sensitivity (for $P[\varepsilon] = 10^{-9}$, a 3 dB power margin, and optimum multiplication) would be 46,1 dBm. In comparison to the quantum limit, this represents a degradation of about 17 dB.

Example 6.6. Let us assume a PIN-diode receiver so that M = F = 1. A simpler approximate formula can then be derived for E_s or P_s because we can assume $P_z \ll P_{cr}$ in (6.43). The following result is obtained

$$E_s = P_s \cdot \tau_T \simeq 2\sqrt{2\kappa k T C_i}\,\frac{hc}{e\eta\lambda} \quad . \tag{6.55}$$

Assuming the same parameters as in Example 6.5 and T = 290 K, we obtain $E_s \approx 18$ fJ or -108 dB(mJ), which is a degradation of about 32 dB from the quantum limit.

Our next task is to consider inter-symbol interference which, as mentioned earlier, is a consequence of the partial overlap of neighbouring pulses. This effect can be simply visualized by observing the input of the decision circuit on an oscilloscope that is triggered by the receiver clock. Then, due to the random occurance of zeroes and ones in the observed time slot, an "eye pattern" such as the one in Fig.6.8a is obtained. An idealized version of the eye pattern is shown in Fig.6.8b from which we see that the eye "opening" is an indication of the noise margin, while the width of the eye shows the timing margin.

From an analytical viewpoint, inter-symbol interference is most conveniently modelled by using a discrete-time formulation. Thus, we write the detector current in the form:

$$i(t) = \sum_{m=-K}^{K} s_m i_s(t - t_m) \quad , \tag{6.56}$$

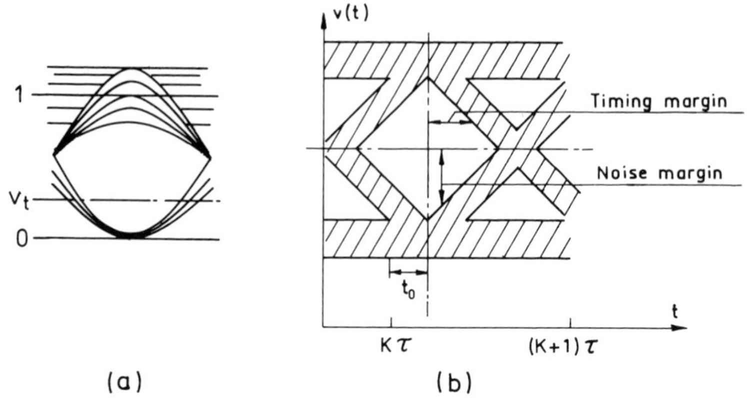

Fig. 6.8. (a) Illustration of a typical eye pattern observed at the input of the decision circuit. (b) An idealized eye pattern showing timing and noise margins

where $s_m = 1$ or 0 (binary case) and $i_s(t)$ is the waveform of a single signal pulse. Using (6.17) the output sample w_ℓ becomes

$$\begin{aligned} w_\ell &= \int_{-\infty}^{\infty} h_R(t_\ell - t')i(t')\,dt' \\ &= \sum_{m=-K}^{K} s_m \int_{-\infty}^{\infty} h_R(t_\ell - t')i_s(t' - t_m)\,dt' \\ &= \sum_{m=-K}^{K} s_m v_{\ell-m} \quad , \end{aligned} \tag{6.57}$$

where $v_\ell = v(t_\ell)$. The complete model requires the addition of noise samples n_ℓ, whose signal dependent variance is given by (6.21). Equation (6.57) then becomes

$$w_\ell = \sum_{m=-K}^{K} s_m v_{\ell-m} + n_\ell \quad . \tag{6.58}$$

For $\ell = 0$ the main component is assumed to be v_0, so that

$$w_0 = s_0 v_0 + \sum_{m\neq 0} s_m v_{\ell-m} + n_\ell \quad . \tag{6.59}$$

Here the term $s_0 v_0$ corresponds to the correct signal value s_0, the second term represents inter-symbol interference, and the third term denotes noise.

The total contribution of the inter-symbol interference can be estimated from the quantity [6.19]

$$e(\underline{s}') = \sum_{\substack{m=-K \\ m\neq 0}}^{K} \frac{s_m v_{-m}}{a v_0} \quad , \tag{6.60}$$

so that (6.59) becomes

$$w_0 = [s_0 + ae(\underline{s}')]v_0 + n_0 \quad . \tag{6.61}$$

Here $\underline{s}' = [s_{-m}, \ldots s_{-1}, s_1, \ldots s_m]$ is the signal vector (excluding s_0) while "a" is half the difference between signal levels (e.g. for 0 and 1, a = 0.5). Using these definitions, we can say that the inter-symbol interference $|e|$ must be less than unity for reliable reception. As before, the actual probability of error has to be written in terms of conditional probabilities, and we obtain

$$P(\varepsilon) = \frac{1}{2}\,\overline{\{P[\varepsilon|0, e(\underline{s}')] + P[\varepsilon|1, e(\underline{s}')]\}} \quad , \quad \text{where}$$

$$P[\varepsilon|0, e(\underline{s}')] = P[-a + ae(\underline{s}') + n_k > 0] \quad , \tag{6.62}$$

$$P[\varepsilon|1, e(\underline{s}')] = P[a + ae(\underline{s}') + n_k < 0] \quad .$$

In (6.62), the average (indicated by the "bar") is to be calculated over all possible combinations of the signal vector $\underline{s}'$. Other assumptions are: (I) binary transmission, and (II) a decision threshold halfway between the 0 and 1 levels.

In general, the evaluation of (6.62) is rather difficult, particularly for slowly dying pulse forms. Thus, a simpler upper bound, based on the maximum value $E_m = \max |e(\underline{s}')|$ of inter-symbol interference, is sometimes used. The probability of error then becomes

$$P[\varepsilon] < \max_{\underline{s}'} P[\varepsilon|\underline{s}] = P[|n| > a(1 - E_m)] \quad , \tag{6.63}$$

which for $\sigma_1 = \sigma_0$, gives the result

$$P[\varepsilon] < Q[(\sqrt{\kappa}/2)(1 - E_m)] \quad . \tag{6.64}$$

Unfortunately, any calculation of the error probability requires knowledge of the value and statistics of inter-symbol interference. Such information is rarely available, or entails extensive numerical computations, and we must resort to simplifications. One such simplification is to assume that the receiver impulse response and the incoming pulse both have Gaussian shapes. If the shape of the former is given by

$$h_R(t) = \frac{\exp(-t^2/2\tau_R^2)}{\sqrt{2\pi}\tau_R} \quad , \tag{6.65}$$

then the bandwidths in (6.23) through (6.26) take the form

$$\left.\begin{aligned} (B_1)^{-1} &= 2\sqrt{2\pi}\,\sqrt{\tau_R^2 + \tau_s^2} \quad , \\ (B_1')^{-2} &= 8\pi\tau_R\,\sqrt{2\tau_s^2 + \tau_R^2} \quad , \\ (B_2)^{-1} &= 4\sqrt{\pi}\tau_R \quad , \\ (B_3)^{-3} &= 32\pi^2\,\sqrt{\pi}\tau_R^3/3 \quad , \end{aligned}\right\} \tag{6.66}$$

where τ_R and τ_s are the standard mean-square spreads of the filter impulse response and the received pulse, respectively. Moreover, assuming the preceding and following pulses to be 1's, the level of inter-symbol interference becomes

$$\zeta = 2e(\underline{s}') \approx 2\,\exp[-\tau^2/(\tau_R^2 + \tau_s^2)] \quad . \tag{6.67}$$

Hence, unlike the attenuation-limited case, we now have optical power at the receiver even for a transmitted zero. As a result, the receiver will require a higher value of P_s to achieve a given error rate. The new equation for P_s can be derived in the same way as before, except that, for a transmitted zero, we will now have a received mean proportional to ζP_s and a variance proportional to $(P_{cr} + \zeta P_s)$. Consequently, $P[V|0]$ in (6.30) must be appropriately modified. Carrying the derivation through, we obtain the result:

$$P_s = \left\{P_z(1+\zeta)/4 + \sqrt{P_z[P_{cr}(1-\zeta)^2 + P_z\zeta/4]}\right\}(1-\zeta)^{-2} \quad . \tag{6.68}$$

This agrees with (6.43) for $\zeta = 0$, but goes to infinity for $\zeta \to 1$. Use of (6.67) and (6.68) shows that (6.6) is correctly expressed, i. e. for $\sqrt{\tau_R^2 + \tau_s^2}/\tau = 0{,}25$ the loss in power is about 1 dB. Beyond that, the loss grows very rapidly with increasing τ_s. If τ_s is known as a function of repeater distance, P_s and P_r in (6.5) can be evaluated, and the correct distance can be iteratively found.

In the foregoing, our appraoch has been to accept the existence of inter-symbol interference and to avoid its influence either by shortening the inter-repeater spacing or by increasing the transmitted power. The alternative, which we will briefly discuss, would be to try to minimize or even remove inter-symbol interference, by using an appropriately dimensioned and tuned filter (the equalizer). The intention is to produce an output spectrum which is known to give zero inter-symbol interference. A common choice is the raised-cosine family (Fig.6.9) which can be expressed in the form:

$$\begin{aligned} C(f,\alpha) &= \tau \quad \text{for} \quad 0 \leq |f| \leq f_0(1-\alpha)/2 \quad , \\ &= \frac{\tau}{2}\{1 - \sin[\frac{2\pi\tau}{2\alpha}(f - f_0/2]\} \quad \text{for} \quad \frac{f_0(1-\alpha)}{2} \leq f \leq \frac{f_0(1+\alpha)}{2} \quad , \\ &= 0 \quad \text{for} \quad |f| > \frac{f_0(1+\alpha)}{2} \quad . \end{aligned} \tag{6.69}$$

In the time domain, this corresponds to the waveform

$$c(t/\tau,\alpha) = \frac{\sin(\pi t/\tau)}{\pi t/\tau} \frac{\cos(\alpha\pi t/\tau)}{(1-4\alpha^2 t^2/\tau^2)} \quad , \tag{6.70}$$

which is clearly zero for any non-zero integer value of t/τ. Obviously, the receiver output can also be expressed in the form

$$V(j\omega) = F_T(j\omega)H_c(j\omega)H_R(j\omega)M\eta e\lambda/(hc) \quad , \tag{6.71}$$

so that $H_R(j\omega)$ is simply given by

$$H_R(j\omega) = \frac{C(f,\alpha)}{F_T(j\omega)H_c(j\omega)} \frac{hc}{M\eta e\lambda} \quad . \tag{6.72}$$

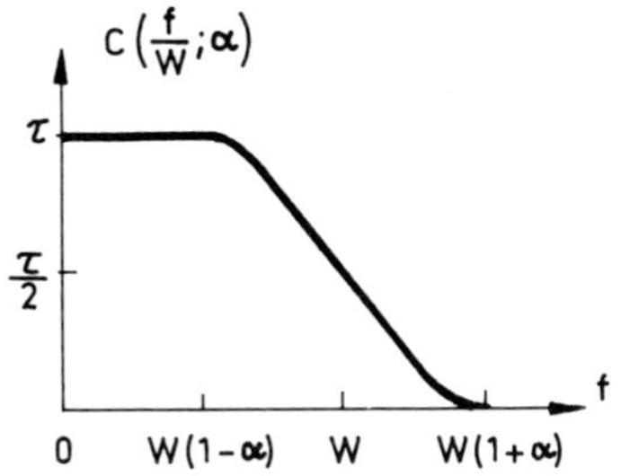

Fig. 6.9. Example of a raised-cosine spectrum ($W = f_0/2$)

Receiver performance with equalization can then be computed from (6.21,22, and 38) after computation of the bandwidths B_1 to B_3. The required transfer function can be approximately implemented by using a sufficiently long delay line.

We conclude this section by observing that the form of linear equalization just discussed necessarily enhances noise at the higher frequencies. Also, for the Gaussian-like transfer function of the optical fiber channel, receiver performance is often degraded by this equalization. Another approach, which we have neglected, is to use non-linear decision feedback [6.20]. This is a technique that can lead to an improvement of several decibels, in effect close to the case of no inter-symbol interference.

6.2 Analog Communication Systems

In the previous section we considered the performance of digital systems, and saw that speech signals can be conveniently and reliably transmitted via PCM channels. Such transmission is not, of course, restricted to speech signals alone, but could, in principle, be used for the transmission of any analog signal, provided that it is first converted into digital form. What do analog systems then have to offer? It would be fair to say that their most important advantageous feature is related to economics, in that analog-to-digital and digital-to-analog conversions are completely avoided, and for multi-channel systems, multiplexing and de-multiplexing costs are significantly lower. The problems are particularly severe for broad-band analog signals, such as those encountered in cable television (CATV) transmission. On the other hand, for signal transmission on-board ships, aircraft, etc., even though the signal bandwidths may be small, the additional cost and complexity of digital transmission is often unacceptable. Thus, for the present, analog systems are a viable alternative for a number of applications, of which video transmission is probably the most interesting.

The simplest configuration is one in which the output electrical signal from a TV camera is used to directly modulate the output intensity of a LED or laser diode [6.21]. The optical signal propagates along the fiber, and is converted back into electrical form at the receiver. Unfortunately, system nonlinearities are a troublesome source of harmonic and intermodulation products that can distort the observed "picture", if the products fall within the bandwidth of interest. Since this bandwidth is relatively large (5.5 MHz), nonlinear distortion can be quite serious.

A solution that avoids this problem is to first use vestigial sideband modulation (VSB) on an electrical "subcarrier" [5.10] that in turn modulates the intensity of the source. If the frequency of the subcarrier is sufficiently high, harmonic and intermodulation products will lie well outside the band of interest, and can be easily removed by filtering. Moreover, VSB is also used for broadcasting TV signals, so that after conversion into electrical form at the receiving end, the signal can be normally "viewed" with a standard TV receiver. In addition, frequency-division multiplexing (FDM) techniques, used in conventional (copper) CATV, can also be directly applied for the transmission of a number of TV channels in the optical fiber [6.22-24]. However, in this case, the total number of channels must be carefully limited, in order to minimize the insertion of the harmonic and intermodulation products of any one channel into any other channel. Similarly, inter-channel intermodulation products must also be taken into account.

The importance of reducing nonlinearities in an analog system should now be obvious. In order to obtain some quantitative basis for design, let us postulate a "softly" nonlinear transfer characteristic, as below

$$P_t = a_1 i_t + a_2 i_t^2 + a_3 i_t^3 \quad , \tag{6.73}$$

where P_t is the transmitted optical power, i_t is the modulating current, while the coefficients a_1, a_2, and a_3 describe the source characteristic. Note that we have implicitly assumed that the nonlinearity can be traced back to the source. If this is not the case, P_t should be interpreted to represent the output voltage or current of the receiver.

For the sake of simplicity, let us also assume that the modulating current is a summation of two sine waves as follows

$$i_t = I_1 \cos\omega_1 t + I_2 \cos\omega_2 t \quad , \tag{6.74}$$

where I_1 and I_2 are the amplitudes of the two sinusoidal components of angular frequency ω_1 and ω_2, respectively. Substitution of (6.74) into (6.73) then yields the following expression

$$\begin{aligned} P_t = {} & (a_2/2)(I_1^2 + I_2^2) \\ & + (a_1 I_1 + 3a_3 I_1^3/4 + \tfrac{3}{2} a_3 I_1 I_2^2) \cos\omega_1 t \\ & + (a_1 I_2 + 3a_3 I_2^3/4 + \tfrac{3}{2} a_3 I_1^2 I_2) \cos\omega_2 t \end{aligned}$$

$$+ (a_2/2)(I_1^2 \cos 2\omega_1 t + I_2^2 \cos 2\omega_2 t)$$

$$+ a_2 I_1 I_2[\cos(\omega_1 + \omega_2)t + \cos(\omega_1 - \omega_2)t]$$

$$+ (a_3/4)(I_1^3 \cos 3\omega_1 t + I_2^3 \cos 3\omega_2 t)$$

$$+ (3a_3/4)I_1^2 I_2[\cos(2\omega_1 + \omega_2)t + \cos(2\omega_1 - \omega_2)t]$$

$$+ (3a_3/4)I_1 I_2^2[\cos(\omega_1 + 2\omega_2)t + \cos(\omega_1 - 2\omega_2)t] \quad . \qquad (6.75)$$

This result shows the existence of second and third harmonics, as well as a number of intermodulation products of sum and difference frequencies. We can now confirm our earlier comments by noting that, in the case of an intensity modulated (IM) video signal, the frequencies $f_1 = \omega_1/2\pi$ and $f_2 = \omega_2/2\pi$ could lie anywhere in the range DC...5.5 MHz, so that all the terms in (6.75) can contribute distortion components to the band of interest. On the other hand, in the case of a VSB carrier, f_1 and f_2 will both be close to the center frequency of the carrier. Since the frequency of the carrier will normally be much higher than 5.5 MHz, it is clear that all the harmonic and intermodulation terms in (6.75) will be sufficiently far away from the band of interest, to allow strong attenuation by filtration. In the multichannel case, we could, for example, fit all channel spectra within the band defined by the lowest subcarrier frequency and its second harmonic. This would allow us to filter out all harmonics, but removal of intermodulation terms would be restricted to sum frequencies only. The residual distortion due to difference frequency terms can be estimated with reference to the $\cos\omega_1 t$ term in (6.75), by setting $I_1 = I_2 = I$, and only using the $a_1 I$ amplitude term. By convention, the result is expressed in terms of the *harmonic distortion* coefficients defined below

$$k_2 = a_2 I/(2a_1) \quad , \quad k_3 = a_3 I^2/(4a_1) \quad . \qquad (6.76)$$

From (6.76), we conclude that the relative intermodulation distortion is $2k_2$ at the lower difference frequency, and $3k_3$ at the higher.

Another detrimental effect of nonlinearities, which is worth noting, arises in the transmission of colour signals, and is caused by variations in the amplitude and phase of the colour subcarrier due to changes in the amplitude of the luminance signal. These variations are described in terms of the "differential gain" and "differential phase" of the system [6.25]. The observable effect is a colour distortion due to the former, and a distortion of the hue due to the latter. The existence of differential gain can be easily confirmed from (6.75), in which the ω_1 and ω_2 terms have amplitudes that are functions of both I_1 and I_2. The existence of differential phase is less obvious, and arises from the fact that, in general, the coefficients a_1, a_2, and a_3 in (6.73) are functions of drive-amplitude-dependent phase. As a result, the cosine terms in (6.75) should also contain phase terms.

In the rest of this section, we will assume that the effects of nonlinearities have been reduced to tolerable levels during preliminary design, and that the influence of fiber dispersion is negligible. The latter may be justified by observing that, in an analog system, the role of dispersion is to cause loss at the higher modulation frequencies, and that some loss can be accepted because of the possibility of pre-emphasis or equilization at the transmitter. Of course, it is true that delay distortion also exists for multiplexed signals, but its effect is negligible within an FDM channel.

Thus, the evaluation of system performance is reduced to the calculation of the receiver sensitivity followed by an iterative computation of the link length using (6.5) in Sect.6.1. However, we should bear in mind that the use of a subcarrier will modify the noise behaviour, because signal-dependent noise will be generated by the subcarrier rather than by the modulating signal. We can account for this feature by writing the current produced by the photo-detector in the following form

$$i_s(t) = I_0 \left[1 + \sum_{\ell=1}^{K} m_\ell s_\ell(t)\right] \quad , \tag{6.77}$$

where I_0 is the average current of the diode, m_ℓ is the modulation index, and $|s_\ell(t)| \leq 1$ represents one of the K normalized carriers. The modulation index m_ℓ should be dimensioned to ensure a total index that is smaller than some critical value m, for which the distortion level is acceptable. In other words,

$$\sum_{\ell=1}^{K} m_\ell \leq m \quad . \tag{6.78}$$

This condition could be satisfied by selecting $m_\ell \leq m/K$.

We are now in a position to define a signal-to-noise ratio for our analog system. However, in contrast to Sect.6.1 in which we defined four separate bandwidths in (6.23) to (6.26), here we will simplify the situation, and define a single noise equivalent bandwidth B_N, that can be obtained using (6.25).

Then, using the spectral power density in (6.14), we have the approximate expression

$$\kappa = \frac{(m_\ell I_{s0})^2}{[MeFI_{s0} + 2kT/R_b + 1.4kT(1 + \omega_\ell^2\tau_i^2)/g_m R_b^2]B_N} \quad . \tag{6.79}$$

As in digital systems, for a bipolar transistor the last term in the denominator of (6.79) should be replaced by the spectral density in (6.15).

We saw in Example 6.3 (Sect.6.1), that a signal-to-noise ratio of 21.1 dB was required for a bit error rate of 10^{-9}. In contrast, the standardized signal-to-noise ratios in point-to-point video transmission vary from 50 ... 57 dB (see e.g. [6.25]) and are even higher at the studio level. Given the performance of typical electronic amplifiers, these figures imply the necessity of rather large signals at the receiver. As a result, the signal-dependent noise contribution tends to be large, so that for high quality transmission, we can select a value of M = 1. Under these conditions, we can define P_z and P_{cr} in (6.27,28) as follows

$$P_z = \kappa B_N (K/m)^2 \frac{hc}{\eta\lambda} \quad , \tag{6.80}$$

$$P_{cr} = \frac{kT}{e} \frac{1.4C_i^2\omega_\ell^2}{g_m} \frac{hc}{e\eta\lambda} \quad . \tag{6.81}$$

and can evaluate the required optical signal power from (6.27).

In order to confirm our previous assertion that a relatively large signal is needed, let us assume that K = 3, m = 0.5, C_i = 10 pF, $f_\ell = \omega_\ell/2\pi$ = 62 MHz, g_m = 10 mS, η = 0.5, λ = 850 nm, and T = 300 K. Moreover, in video transmission the required signal-to-noise ratio is normally specified as a weighted value at the output of the video amplifier of the TV receiver. The weighting corresponds to "white" noise filtration by an RC filter with a time constant of τ_N = 0.33 µs. The corresponding value of B_N can be obtained from (6.25) in Sect.6.1 to be $B_N = 1/(4\tau_N)$, or 760 kHz. If we further stipulate a value of $\kappa = 10^5$, use of (6.80) and (6.81) yields P_z = 1.28 µW and P_{cr} = 160.8 µW. With these values, the required optical power, from (6.27), becomes P_{s0} = 14.95 µW or -18.3 dBm. We see that, to achieve a reasonable link length, a substantial amount of power must be launched into the fiber.

The fact that a large amount of power is available from DH stripe lasers makes these devices very interesting. However, because we would simultaneously like the source to be linear, such diodes should be single mode. Unfortunately, under such conditions, the coherence properties of the source are improved, and we have to contend with an additional source of "noise", known as modal noise [6.26] which appears as an unwanted modulation of the received signal.

The existence of modal noise can be qualitatively understood on the basis of our considerations in Sect.3.7, from which we know that when a multimode fiber is illuminated by a monochromatic source, the near and far-field intensity patterns consist of speckles, provided that the coherence time of the source is more than the inter-modal delays. We should also bear in mind that speckle patterns are rather sensitive to changes in the source wavelength (e.g. via cavity heating) and to mode delay variations in the fiber due to pressure (acoustic waves) and temperature (Sect.6.4).

Consider then some fiber joint that is located at a point where sufficient modal coherence is available for the generation of interference speckles. If this joint is transversely or longitudinally mis-aligned, only some of the speckles of the launching fiber will be collected by the receiving fiber. Let us further assume that the joint itself is "permanent", i.e. the relative displacements remain stationary. Even under these conditions, speckles that are around the periphery of the fiber's acceptance cone, will tend to "jump" in and out of the cone, due to even minute changes in the source wavelength, and perhaps less seriously due to changes in the external temperature and pressure. It is then clear that the power accepted by the receiving fiber will be modulated. Because of the randomness of the factors producing this modulation, the propagating power will seem to contain noise. Indeed, given the large signal-to-noise ratio requirements of the type of video system we have considered in this section, the existence of modal noise is a formidable drawback that cannot be overcome by the simple expedience of increasing the transmitted power.

What is to be done? The problem, in fact, boils down to the avoidance of interference. Three basic approaches are possible: 1) we can use a single-mode fiber, 2) we can locate the first joint well beyond the point at which the inter-modal delays exceed the coherence time of the source, and 3) we can use sources that have short coherence times or are incoherent. The first choice will no doubt become widespread as the technology matures. The second choice is inconvenient both because a "pig-tail" at the source might not be useable, and because of the necessity for long cables on massive bobbins. The third choice implies the use of LEDs, or laser diodes with a broad spectrum, so that material dispersion will tend to be an important factor. It should also be noted that a *perfectly* aligned joint between *identical* fibers would also eliminate modal noise. In practice, we can only achieve low-loss joints and nearly identical fibers, so that a certain amount of residual modal noise can always be expected. Similarly, time-dependent mode conversion between propagating and non-propagating modes will also make a contribution to modal noise. However, for sufficiently low levels, digital PCM systems with their relatively low signal-to-noise ratio requirements, could be used at the expense of greater complexity in implementation.

Finally, we would like to mention that although we have only discussed VSB-IM in any detail, and have referred to IM and PCM, other modulation techniques [6.27] such as pulse-frequency modulation (PFM), pulse-position modulation (PPM), and a host of other well-known techniques and variations could also be used. For a treatment of the conventional modulation techniques, we refer the reader to [6.28].

6.3 Instrumentation and Data Systems

The systems that were considered in Sects.6.1 and 6.2, basically exploited the broad-band and low attenuation features offered by optical fibers. However, instrumentation and data systems neither require particularly low attenuations (typical links' lengths are just tens to hundreds of meters) nor specially large bandwidths (data rates range from a few kilobits to, at most, some megabits per second). In fact, from the viewpoint of these systems, the most useful fiber properties are: 1) that fibers are immune to electromagnetic interference, 2) that they are small and light, and 3) that they do not constitute a fire or explosion hazard.

The simplest application configuration is one in which information is transferred from one point, e.g. an instrument or computer terminal, to another point, e.g. the central computer. The information flow could be one way (simplex), two way but with the participants taking turns (half duplex), or unrestrictedly two way (full duplex). An extension of this type of system is the multi-point or party-line configuration, in which a number of parties are allowed access into the line, and can communicate with the central station using some line discipline such as polling. In hub polling, for example, the central computer sequentially interrogates each secondary station for its data. The multi-point configuration does not normally allow communication amongst the secondary stations or terminals.

At a higher level of complexity, we have systems which also allow interterminal communication, and use topologies such as the star or the loop. In the former, each terminal has an independent link to the central station, while in the latter, a single line visits each terminal and comes back to its beginning. In the loop system, all information necessarily circulates (together with the address of its destination) and is picked up by the addressed terminal. However, in the star configuration this is a matter of choice and is determined by the type or types of couplers used. If a star coupler or repeater is used, rather flexible two way communication is possible, either between terminals or between the terminal and the central computer. It should be noted that more complex geometries can also be obtained by a combination of the above topologies.

For any given application, the choice of topology is mostly determined by two conflicting factors: 1) the required length of fiber and 2) the desired level of reliability. Thus, for example, the star requires substantial lengths of fibers, but is very reliable, particularly if the center point is secured e.g. by doubling it. In contrast, multi-point and loop systems require less fiber, but can be easily "muted" by even a single fault near the main station.

With the foregoing background information, it is now easy to understand that, except for the simple point-to-point system, all the other configurations require efficient couplers for the insertion and extraction of data. Multipoint and loop topologies require the conceptually simple T-coupler, while the star network, as mentioned earlier, may require a star coupler. A number of techniques are available for constructing these [6.29-35], some typical ones being listed below.

1) *Active Repeater Branching*. In this method, incoming fibers are directly butted against a broad-area detector, while out-going fibers are all simultaneously excited by a broad-area LED. Optical re-transmission may be with or without signal regeneration. Obviously, both T and star couplers can be made using this technique, and no optical power loss is incurred. However, the penalty is a potential reduction in reliability due to the use of active elements.

2) *Optical T-Couplers*. These can be formed by conventional beam splitting techniques. A simple variation is to twist together two plastic clad fibers and to melt the cladding until the desired amount of coupling is achieved. In both cases, an inherent division loss, e.g. 3 dB per branch, is inevitable.

3) *Passive Star Couplers*. These are formed by butting the incoming fibers against a cladded mixing rod which evenly "blends" all the signals. Branching can then be simply obtained by butting the out-going fibers against the oppo-

site end of the mixing rod. Another variation is to terminate the mixing rod in a mirror which returns the "blended" signals for re-coupling. The coupling efficiency can be maximized by maximizing the "packing fraction" of fibers so that the maximum amount of power would be incident on the fiber cores. However, as in the passive T-coupler case, division loss is always incurred. For example, in a 7 fiber star, the division loss is 10 lg 7 = 8.5 dB.

Having selected a suitable topology for our data or instrumentation system, we must next examine the type of electronic interface which is used. For data systems, interfaces such as the V. 10, V. 11, and V. 24 have been defined [6.36], while for instrumentation systems the IEEE 488 standard digital interface is available [6.37]. The data interfaces provide signals in a serial format, while the IEEE interface bus provides 8 bit data bytes in parallel format. A common feature of both data and instrumentation interfaces, is the necessity for numerous control and "hand-shake" lines that necessarily increase the cable size in conventional systems. However, an optical fiber, being wide-band, could easily accept the serial transmission of both data and control signals. Of course, it is true that the simplicity of control using separate lines would have to be replaced by complex protocols [6.38] requiring sophisticated link controllers, but in many applications this would be offset by the advantages of fiber-optics.

The actual design of the system can be achieved in much the same way as in Sect.6.1, except for the relatively low modulation rates and short link distances. As a result, the dispersion and attenuation characteristics of the fiber are rarely a problem. Hence, light-emitting diodes are usually adequate for the generation of light, while PIN diodes or even photo-transistors can be used for detection.

The most serious design problem is related to the source-to-fiber coupling efficiency, which can be increased by the use of step-index fibers of relatively large numberical aperture and core diameter. Alternatively, fiber bundles can be used to improve the source-to-fiber area match, and to increase reliability by utilizing the inherent channel redundancy in the bundle. Of course, as the coupling efficiency is increased, fiber loss will once again become significant. In the case of a compromise, we could define the following figure-of-merit, as the quantity to be maximized:

$$F_M = A_N^2 \exp(-\alpha L) \quad , \tag{6.82}$$

where A_N is the numerical aperture of the fiber, L is its length, and α is its attenuation per unit length.

In concluding this section, we should say a few words about line codes suitable for instrumentation and data systems. Evidently, we could simply adopt one of the codes discussed in Sect.6.1. However, it may often be more economical to use a code that has been specifically designed for short haul systems. An example of one such type is the transition code, in which constant signal levels are represented by the average optical output P, while "01"s are represented by 2P and "10"s by zero optical power. In addition, long transition-free periods are avoided by transmitting violations at pre-set intervals [6.39-41].

6.4 Optical Fiber Sensors

Optical fiber sensors represent a relatively new and fast growing family of optical devices which can be used for the measurement of physical parameters such as temperature and pressure. The most commonly used technique is to "probe" the parameter of interest with a coherently illuminated fiber and to compare the output with that of a "shielded" reference fiber. It is clear that the external influence can change the phase of the light in the probe by changing the fiber length, the fiber core diameter, or the refractive index. We can express the resultant phase change $\Delta\phi$ in the following manner

$$\Delta\phi = \beta(\Delta L) + L(\Delta\beta) \quad , \tag{6.83}$$

where L is the length of the fiber, β is the axial propagation constant, and (ΔL) and $(\Delta\beta)$ are the changes in L and β, respectively. For pressure measurements the first term in (6.83) represents the physical change in the length of the fiber due to strain [6.42]. In other words, $\beta\Delta L$ can be written in the form

$$\beta\Delta L = -\beta(1 - 2\mu)\,\frac{L(\Delta P)}{E} \quad , \tag{6.84}$$

where (ΔP) is the change in applied pressure, μ is Poisson's ratio, and E is Young's modulus.

The second term in (6.38) is somewhat more complicated, and includes possible changes in diameter and refractive index. Namely,

$$L\Delta\beta = L\,\frac{\partial\beta}{\partial n}\,(\Delta n) + L\,\frac{\partial\beta}{\partial a}\,(\Delta a) \quad , \tag{6.85}$$

where n is the core refractive index, (Δn) its change, and (Δa) is the change in radius. For a weakly guiding fiber (Chap.3), core and cladding refractive indices are nearly equal, particularly in a single mode fiber. Thus, β can be approximated by (nk_0) and $\partial\beta/\partial n$ by $\partial(nk_0)/\partial n$. It can also be shown that the refractive index change due to strain is approximately given by [6.42]

$$\Delta n = \frac{1}{2} n^3 (\Delta P/E)(1 - 2\mu)(2P_{12} + P_{11}) \quad , \tag{6.86}$$

where P_{11} and P_{12} are the relevant components of the strain-optic tensor (Sect.2.2). Moreover, the second term in (6.85) can be shown to be negligible for the weakly guiding fiber [6.42], so that the pressure sensitivity of the fiber becomes

$$\frac{\Delta\phi}{L \cdot \Delta P} = \frac{-\beta(1 - 2\mu)}{E} \left[1 - \frac{n^2}{2} (2P_{12} + P_{11})\right] \quad . \tag{6.87}$$

For fused silica, this typically corresponds to about -4.1×10^{-5} rad/(Pa · m). Thus, if such a fiber would be used as the probe in a two-arm fiber interferometer (Fig.6.10), with the other arm isolated from external influence, the one-fringe displacement would require about 153 [kPa · m]. This sensitivity can be substantially different for multi-component glasses because of their wide range of physical parameters.

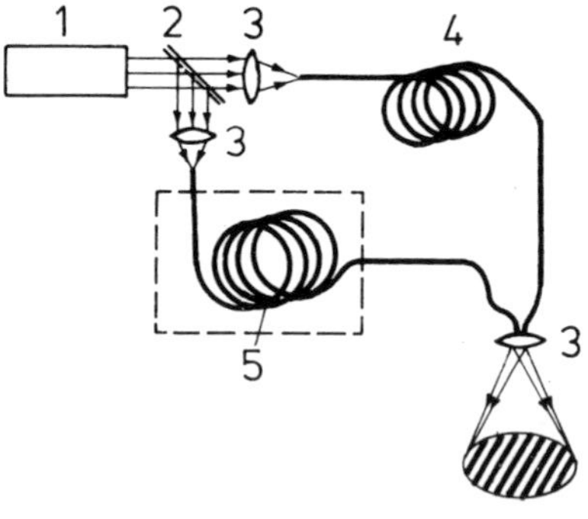

Fig. 6.10. Illustration of the two-arm interferometer

Similar calculations can also be made for the temperature sensitivity of the fiber, and a value of about 107 rad/(°C · m) is obtained for fused silica [6.42]. Moreover, experimental investigations, both for pressure and temperature sensitivities, indicate reasonable agreement with these figures [6.42,43], but also bring forth the importance of certain factors that were thought to be of secondary importance. For example, the plastic coating of the fiber can dramatically enhance the pressure sensitivity, instead of damping the acoustic wave [6.43]. The reason seems to be that the plastic elongates much more than the bare fiber, and forces the fiber to stretch with it.

An interesting application of the pressure sensing fiber is to use it to form a hydro-phone. The performance of such hydro-phones has been analyzed in [6.43-45], and it has been shown that the human threshold of hearing can be surpassed using about 30 m of fiber, while the performance of conventional high quality hydro-phones can be equalled using about 1 km of fiber. Experimental results (Fig.6.11) indicate that these predictions are pessimistic for plastic-coated fibers, and that even 10 m may be sufficient [6.43].

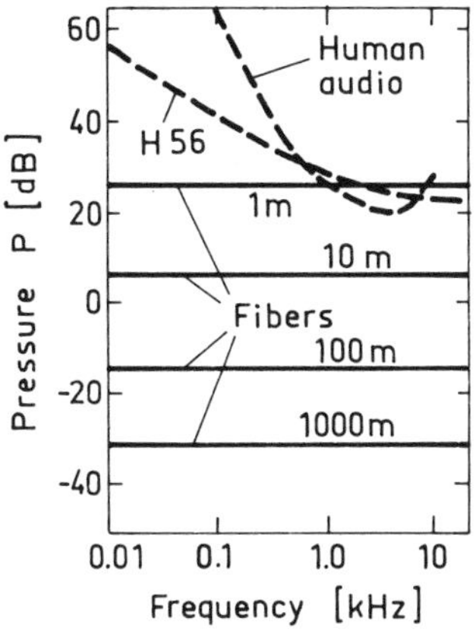

Fig. 6.11. Performance of some fiber Hydrophones (reference pressure 1μ Pa) [6.43]

Fig. 6.12. (a) The classical Sagnac "ring" interferometer. (b) The fiber equivalent of the above ▼

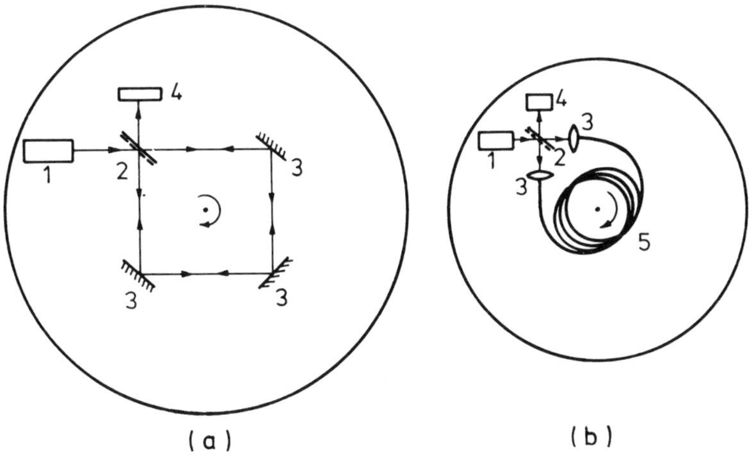

The gyroscope is another type of sensor which can be based on fiber-optics. In principle, the device is the Sagnac "ring" interferometer, as in Fig.6.12a, with the conventional air paths replaced by single-mode fibers, as shown in Fig.6.12b. It is easy to understand that if the platform is rotated, the clockwise and counter-clockwise beams take different times to complete the loop. For a circular loop of fiber placed symmetrically about the axis of rotation, the time difference between clockwise and counter-clockwise paths is easily shown to be

$$\Delta t \simeq \frac{\Omega A^2 N}{\pi c^2} \quad , \tag{6.88}$$

where A is the perimeter of the loop, Ω is the angular velocity of rotation, c is the velocity in the fiber, and N is the number of fiber turns in the loop. The corresponding phase shift is obviously given by

$$\Delta\phi = \frac{(2\pi c)}{\lambda} \Delta t = \frac{2\Omega A^2 N}{\lambda c} \quad , \tag{6.89}$$

where λ is the free-space wavelength. We see that the angular velocity is directly proportional to the phase shift. Moreover, the sensitivity can be increased by increasing the number of loops N, and even more dramatically by increasing the diameter (and hence the perimeter A) of the loop. For detailed sensitivity analyses, we refer the reader to [6.46,47].

6.5 Wavelength-Division Multiplexing

In our discussion on PCM systems in Sect.6.1, we had mentioned that a number of channels are normally time-division multiplexed (TDM), in order to use time "slots" which would otherwise be empty. The other well-known technique for increasing the information-carrying capacity of a channel is to frequency-division multiplex (FDM) the signals to be transmitted, and hence fill up empty frequency bands. For example, we saw in Sect.6.2 that a number of TV channels could be conveniently transmitted in this way. In optical fiber communications, besides TDM and FDM, we can also increase the channel capacity by wavelength-division-multiplexing (WDM).

This is the technique whereby a number of optical "carriers" at different wavelengths are allowed to propagate within the fiber, with each carrier further multiplexed in the time or frequency domain. Obviously, to extract the maximum benefit from WDM, we require optical sources that have narrow spectra as well as closely spaced central wavelengths, because in this way, the largest possible number of carriers can be fitted into a given transmission "window" of the fiber. Narrow spectra are, of course, needed to prevent excessive inter-channel interference (cross-talk). An additional constraint is related to the wavelength stability of the source, because any drift in the central wavelength will also lead to cross-talk. In the presence of such drift, we must compromise on the total number of WDM channels. This is the situation we might face when considering the use of laser diodes, because, although these have a relatively narrow spectrum, they are rather sensitive to temperature change. On the other hand, lightemitting diodes are more stable but tend to have broad spectra.

The other key factor in efficient implementation of WDM is related to the availability of suitable multiplexing and de-multiplexing devices: from the former we require the minimum possible insertion loss, while from the latter we additionally require high wavelength selectivity. If interference filters are used, a single device often serves both purposes, in which case the two requirements will obviously merge. However, in principle, wavelength selectivity is not necessary for multiplexing, because we could, for example, use the non-selective branching couplers considered in Sect.6.3, even though a high insertion loss would then be incurred.

In any case, we see that the problems encountered in WDM are mostly technological rather than conceptual. As such, the rest of this section will be a brief look at some of the methods used for the implementation of multiplexers and de-multiplexers. Two basic phenomena are presently used: 1) the wavelength dependence of the angle of reflection or transmission in angularly dispersive devices, such as prisms or diffraction grating, and 2) the wavelength dependence of the transmission and reflection coefficients of multilayer interference filters.

An example of a grating de-multiplexer [6.48] is shown in Fig.6.13. Input and output fibers are arranged such that they form a linear array with fixed inter-core separations. A blazed grating is then used to direct the various maximum intensity orders onto the required fiber. The advantage of this type of grating is that, by choosing the correct blaze angle, we can not only determine the order in which we have maximum intensity at a given wavelength, but we can also suppress the uninteresting zero-order diffraction image [6.49]. Energy which would otherwise be wasted in the zero-order image is thus conserved and increases the intensity in the chosen order. The required precision is attainable by the anisotropic etching of silicon wafers by using

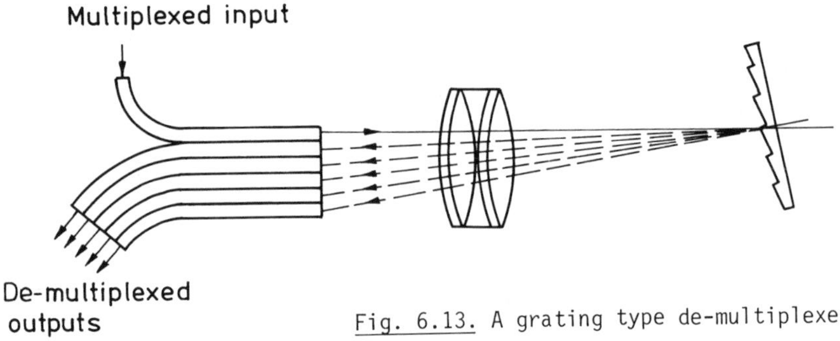

Fig. 6.13. A grating type de-multiplexer [6.48]

photolithographic techniques [6.48]. As a result, insertion loss figures of less than 2 dB can be achieved. However, this method suffers from the common draw-back of most angularly dispersive devices, namely that they are not very useful for broad spectral sources such as LEDs.

This is one of the reasons for the strong interest in devices based on the interference filter principle [6.48], besides the fact that they can be easily arranged to work both as multiplexers and de-multiplexers. An interesting device based on this principle has been described by HASHIMOTO and NOSU [6.50], and is shown in Fig.6.14. The structure consists of 8 different interference filters, 4 on each side of the glass substrate. Each filter is transparent to only one wavelength and reflects all others. In the multiplexer operation each wavelength is brought to the device by a fiber whose output is collimated by a rod-lens assembly (not shown) of variable focal length. The collimated beam corresponding to the input λ_1 passes through its own interference filter and is re-focused to the output fiber. All other λ inputs undergo one

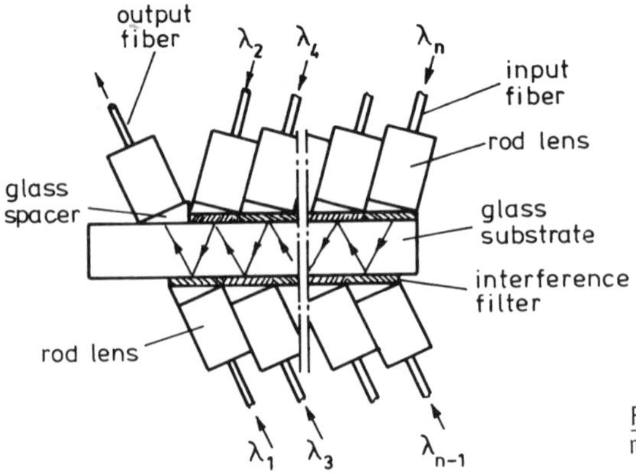

Fig. 6.14. An interferometer type multi/de-multiplexer [6.50]

or more reflections, as shown by the zig-zag ray path in Fig.6.14, and are eventually re-focused onto the output fiber. In the de-multiplexer operation the "output" fiber becomes the input fiber while all the other fibers provide the output wavelengths of interest. In practice these could be replaced by detector diodes in order to reduce coupling losses. Using this approach the sum of multiplexing and de-multiplexing losses easily remain under 3 dB [6.50] and can be expected to improve with advances in interference filter technology.

Having briefly discussed this interesting topic, we now conclude this section by reminding the reader that the need for WDM can also be simply bypassed by increasing the number of fibers in a cable. However, such a solution may not be cost-effective, particularly when cable and installation costs are high. On the other hand, WDM could potentially yield a manifold increase in capacity, especially if the wavelength setting and jitter of laser diodes would be tightly controlled.

Selected Problems

The problems presented in the following have been compiled as far as possible, in the same order as the material in the book. The purpose of the problems is to provide both a "feeling" for the parameters of optical fiber systems, and a deeper insight of some of the techniques. For problems of the latter type, reference to the cited works is strongly recommended. Moreover, for some of the problems requiring numerical solutions, we suggest the use of a programmable calculator or a computer terminal.

1) Using the two-dimensional form of Maxwell's equations and assuming the incidence of a plane wave at a plane interface, show the validity of the law of reflection and refraction (Snell's law), and calculate Fresnel's formulae as in (3.3) and (3.4).

2) Using (3.22), calculate the depth (1/e point) to which power penetrates into the rarer medium. Assume $n_1 = 1.5$ and n_2 to be 1% smaller. Plot the penetration depth as a function of θ_i. Assume distances to be in λ units.

3) Assuming the planar step-index waveguide structure of Fig.3.21 and using (3.20), (3.21), and (3.43), derive the eigenvalue equation for TM modes.

4) Derive the cut-off V value for TE and TM modes of index m and also for the case of $n_c = n_s$. Take $V = kd\sqrt{n_f^2 - n_c^2}$. Calculate the number of TE and TM modes for a structure with $n_f = 1.540$, $n_c = 1$, $n_s = 1.530$, $d = 10\ \mu m$, and $\lambda = 850$ nm. (As in Fig.3.23, assume that the phase shift becomes zero at cut-off).

5) In the paraxial approximation ($ds \approx dz$), the eikonal equation (3.52) becomes

$$n(x)\,\frac{d^2x}{dz^2} = \frac{dn(x)}{dx} \quad .$$

Assuming the linearly graded profile

$$n(x) = n(0)(1 - \Delta|x/a|) \quad ,$$

derive the ray path in the planar waveguide and evaluate the period of the ray.

6) Derive (3.58) from (3.57) for constant n(r). Then, assuming $\underline{E} = \underline{E}_0 \exp[jk_0 s(\underline{r})]$, show that (3.52) is obtained.

7) Show that the eikonal equation is equivalent to Hamilton's equations:

$$\frac{\partial x}{\partial z} = \frac{k_x}{k_z} \quad ,$$

$$\frac{\partial k_x}{\partial z} = \frac{1}{2k_z} \frac{\partial^2 k(x)}{\partial x^2} \quad .$$

Solve these equations for a parabolic index profile and derive the ray period $z(k_z)$ in the non-paraxial case. Calculate $\Delta z = z_{max} - z_{min}$ for $k_{z_{min}} = k_0 \cos\theta_{ic}$.

8) Solve the above equations for the linear profile of Problem 5 and calculate $z(k_z)$ and Δz.

9) Derive the formula for the numerical aperture of a step-index fiber in terms of the core and cladding refractive indices. Calculate the numerical aperture for $n_1 = 1.5$ and a 1% core-cladding difference. What is the angle of the external acceptance cone? (Assume meridional rays only).

10) Estimate the width of the impulse response of a step-index fiber using ray optics. Assume an evenly excited fiber end, and use the minimum and maximum meridional ray paths. Assuming the parameters in Problem 9, what would be the bandwidth of such a fiber? ($B = 0.35/\Delta t$).

11) The core diameter of a certain step-index fiber is 100 μm, and the thickness of its cladding is 50 μm. If its numerical aperture is 0.2 and the core index is 1.5, calculate the fiber's V parameter at λ = 850 nm and 1300 nm, and the number of guided modes. How much should the diameter be reduced for single mode operation?

12) Starting from (3.91), show that (3.93) and (3.95) follow.

13) Derive (3.99) from Maxwell's time-independent curl equations. Comment on the validity of (3.99) for different index profile.

14) Satisfy the boundary conditions to evaluate the constants in (3.97). Show that (3.100) and (3.101) follow. (N.B. This derivation involves some lengthy algebra and a degree of care is required).

15) Show that (3.101) leads to TE and TM fields for $\nu = 0$.

16) Show that in the weakly guiding approximation (3.106), $HE_{\nu\mu}$ modes are degenerate with $EH_{\nu-2,\mu}$ modes. (Use the recursive relations for Bessel functions).

17) Calculate the propagation delay $\tau_{\mu\nu}$ of mode $\mu\nu$ in a step-index fiber, by taking the derivative of the WKB eigenvalue equation (3.123) with respect to k_0. If phase shifts upon reflection are taken into account (always π for graded-index profiles), the following eigenvalue equation is obtained:

$$\mu = 1/4 + (1/\pi)\left[\sqrt{u^2 - \nu^2} - \nu \operatorname{arc\,cos}(\nu/u) - \operatorname{arc\,cos}\left(\frac{\sqrt{u^2 - \nu^2}}{V}\right)\right] .$$

Derive an expression for the propagation delay $\tau_{\mu\nu}$ using this relationship.

18) Derive (3.122) from (3.119) and (3.121).

19) Plot the function under the integral sign in (3.123) as in Fig.3.37 for various values of ν, assuming the following parameters: $\lambda = 850$ nm, $n(0) = 1.5$, $\Delta = 0.01$, and $a = 100$ μm.

20) Equation (3.150) gives an expression for the normalized time delay t_δ within the α-profile approximation. If only the first-order term in δ is considered, the optimum value of α is $\alpha_0 = 2 + \zeta$. Evaluate the optimum α profile when (a) the first- and second-order terms are included, and (b) when all three terms in (3.150) are included.

21) Equating ζ to zero in (3.154), draw the impulse response and evaluate the pulse width in units of ns/km. Then, calculate the frequency response of the fiber by fitting the exponential function $A[1 - \exp(-t/\tau)]$ to the curved part of the impulse response. What is the 3 dB bandwidth of the fiber? Comment on the dimensions of the 3 dB bandwidth.

22) Derive (3.152) and (3.153) from (3.150) and (3.151).

23) The coupling efficiency given in (4.43) pre-supposes a perfect Lambertian source. Assuming the source radiance to be $B(\theta) = B_0 \cos^g\theta$, re-derive the coupling efficiency and compare it to (4.43).

24) For multimode fibers, describe some mechanisms that can cause departures from the exponential power decay law of (5.6).

25) Explain the dispersion mechanisms in single-mode fibers, as well as their contribution to the total dispersion at different wavelengths.

26) Derive the leaky-mode correction factor expression in (5.39) by starting from (5.36).

27) Derive (5.52) and (5.54) from (5.42) and (5.45).

28) Point out the leaky-mode regime of a parabolic fiber in the (u, ν) plane. In which region are the weakly leaky modes located? (Assume $n(0) = 1.46$, $\lambda = 850$ nm, $a = 20$ μm, and $\Delta = 0.027$).

29) In the above, if we do not make the assumption of a continuous modal distribution but consider discrete modes, then u and ν become

$$\nu = 0, 1, 2, 3, \ldots \quad ,$$

$$u = (w_0/a)\sqrt{m} \quad , \quad m = 1, 2, 3, \ldots \quad ,$$

where

$$w_0^2 = 2a/[n(0)k_0\sqrt{2\Delta}] \quad ,$$

and

$$m = 2\mu + \nu + 1 \quad .$$

Show that the latter result is valid, and mark the allowed (discrete) leaky modes in your (u, ν) diagram. Estimate the number of weakly leaky modes (attenuated by less than 50% after 1 m) in the fiber and mark them in your diagram. You may compute the attenuation factor from (5.55). Note also that $\beta^2 = n^2(0)k_0^2 - 4m/w_0^2$ in a parabolic fiber.

30) The numerical aperture of a multimode fiber may be defined via the far-field pattern of (5.57).

a) Draw the above pattern for $n(0) = 1.5$, $\Delta = 0.01$, and $\alpha = 1, 2, 3$, and ∞.

b) Design a fiber such that 90% of the optical power is contained within a numerical aperture of 0.2, or within the half angle $\theta \leq \arcsin 0.2$.

31) An optical PCM system uses a 1B2B-type line code, and is based on a fiber cable with the following characteristics:

- Loss: 3 dB/km ,
- Splice spacing: 1.5 km ,
- Splice loss: 0.2 dB ,
- Critical length: $L_c = 1$ km ,
- $\sigma_{TOT} = 1$ ns/km (laser diode) ,
- $\sigma_{chr} = 5$ ns/km (LED) .

The allowed level difference between input and output is 50 dB irrespective of line rate.

a) Evaluate the critical modulation rate f_{cr} both for the LED and for the laser diode.

b) Compute the curves $L = F(f_m)$ for both sources, and mark the regions of power-limited and dispersion-limited transmission.

c) In particular, evaluate the repeater spacings for interface bit rates of 2.048, 8.448, and 34.368 Mbit/s.

32) Calculation of power budget.

a) Compose the power budget for the above as in Table 6.2, assuming a bit rate of 8.448 Mbit/s, a 5 dB impairment allowance, and an 18 dB coupling loss for the LED.

b) Repeat for a laser diode assuming a 60 dB allowed level difference between input and output, and a coupling loss of 5 dB.

33) The critical modulation rate in (6.10) does not account for mode mixing. Develop a new formula using (6.8) with p = 0.8, and assuming some value for σ_{chr}.

34) Compose a table that compares the efficiencies of the mBnB types of line codes, for n > m and n = 1...8 . The efficiency is measured by using a bandwidth expansion factor

$$k_{exp} = f_m/R \quad ,$$

where f_m is the modulation rate, and R is the numerical (bit) rate.

35) In order to realize the circuits or functions for encoding and decoding, line codes have to be formally defined. For some codes like AMI and its derivatives, a state concept also needs to be defined. Thus the encoder and decoder are modelled as finite state automata. Develop the encoding equations

$$[v_{2k} \;,\; v_{2k+1} \;;\; s_{k+1}] = F[u_k \;;\; s_t]$$

and the decoding equations

$$[\hat{u}_k \quad , \quad e_k] = G[v_{2k} \quad , v_{2k+1}] \quad ,$$

where $\{u_k\}$ are the information bits, $\{v_k\}$ the line code bits, $\{\hat{u}_k\}$ the decoded information bits, and $\{e_k\}$ the error indicator bits. Possibly using some literature on digital logic design, develop these equations for the line codes

a) 2-level AMI ,

b) CMI ,

c) Bi-phase .

36) The 2-level AMI line code may be decoded by using the following "correlation" technique:

- add the encoded digital signal to its half-symbol-delayed version, to obtain the 3-level AMI signal;
- decode the 3-level AMI as usual (by rectification).

a) Show by an example (or even analytically) that the method works.

b) Draw the block diagram of an AMI-detector that operates according to the above principle.

37) Evaluate individually the components of the noise-spectral power density given in (6.14), by assuming the following parameters:

- optical input power to detector: $P_i = 1$ nW ,
- optical wavelength: $\lambda = 0.85$ μm ,
- detector quantum efficiency: $\eta = 0.8$,
- avalanche photodiode multiplication: $M = 100$,
- avalanche ionization ratio: $k' = 0.02$,
- detector circuit parallel resistance: $R_b = 1$ MΩ ,
- detector circuit capacitance: $C_i = 2$ pF ,
- preamplifier FET transconductance: 6.5 mS .

Draw asymptotic curves for the following contributions by expressing the spectral density in dB with respect to 1 A^2s, and by using a logarithmic frequency scale.

a) signal-dependent shot noise,

b) thermal noise,

c) preamplifier shot noise.

In particular, note the frequencies at which two of the three contributions are equal.

38) If the mean-square value of the equivalent voltage generator in Fig.6.5 is given by the well-known relationship

$$\overline{e_n^2} = 4\,kT\Delta f(0.7/g_m) \quad ,$$

show that the equivalent current generator in the figure has the last term in (6.14) for its noise-spectral density.

References

1.1 E. Risberg: *A History of the Finnish Telegraph Administration 1855-1955* (General Direction of Posts & Telecommunications, Helsinki 1955) (In Finnish)

1.2 R. Kompfner: Appl. Opt. *11*, 2412 (1972)

1.3 D. Hondros, P. Debye: Ann. Phys. *32*, 465 (1910)

1.4 N.S. Kapany: *Fiber Optics Principles and Applications* (Academic, New York 1967)

1.5 K.C. Kao, G.A. Hockham: Proc. IEE *113*, 1151 (1966)

1.6 F.P. Kapron, D.B. Keck, R.D. Maurer: Appl. Phys. Lett. *17*, 423 (1970)

1.7 D. Gloge (ed.): *Optical Fiber Technology* (IEEE Press, New York 1976)

1.8 M.K. Barnoski (ed.): *Fundamentals of Optical Fiber Communications* (Academic, New York 1976)

1.9 H.-G. Unger: *Planar Optical Waveguides and Fibers* (Clarendon, Oxford 1977)

1.10 J.E. Midwinter: *Optical Fibers for Transmission* (Wiley, New York 1979)

1.11 S.E. Miller, A.G. Chynoweth (eds.): *Optical Fiber Telecommunications* (Academic, New York 1979)

1.12 H. Kressel (ed.): *Semiconductor Devices for Optical Communications*, Topics in Applied Physics, vol 39 (Springer, Berlin, Heidelberg, New York 1980)

2.1 J.M. Ziman: *Principles of the Theory of Solids* (Cambridge University Press, Cambridge 1964)
O. Madelung: *Introduction to Solid-State Theory*, Springer Series in Solid-State Physics, Vol.2 (Springer, Berlin, Heidelberg, New York 1978)

2.2 J.P. McKelvey: *Solid State and Semiconductor Physics* (Harper and Row, New York; J. Weatherhill, Inc., Tokyo 1966)

2.3 S.M. Sze: *Physics of Semiconductor Devices* (Wiley, New York 1969)

2.4 H. Kressel, J.K. Butler: *Semiconductor Lasers and Heterojunction LEDS* (Academic, New York 1977)

2.5 A. Einstein: Phys. Z. *18*, 121 (1917)

2.6 F.T. Arecchi, E.O. Schulz-Dubois: *Laser Handbook* (North-Holland, Amsterdam; Elsevier, New York 1972)

2.7 M. Young: *Optics and Lasers*, Springer Series in Optical Sciences, Vol.5 (Springer, Berlin, Heidelberg, New York 1977)

2.8 P.K. Tien: Rev. Mod. Phys. *49*, 2 (1977)

2.9 I.P. Kaminow: IEEE Trans. MTT-*23*, 57 (1975)

2.10 J.M. Hammer: "Modulation and Switching of Light in Dielectric Waveguides" in *Integrated Optics*, 2nd ed., ed. by T. Tamir, Topics in Applied Physics, Vol.7 (Springer, Berlin, Heidelberg, New York 1979)

2.11 M. Born, E. Wolf: *Principles of Optics* (Pergamon, Oxford, New York 1975)

2.12 A. Yariv: *Introduction to Optical Electronics* (Holt, Rinehart, and Winston, New York 1971)

2.13 I.P. Kaminow, I.R. Carruthers, E.H. Turner, L.W. Stulz: Appl. Phys. Lett. *22*, 540 (1973)

2.14 I.P. Kaminow, L.W. Stulz, E.H. Turner: Appl. Phys. Lett. *27*, 555 (1975)
2.15 D.A. Pinnow: IEEE J. QE-*6*, 223 (1970)
2.16 R.J. Pressley (ed.): *Handbook of Lasers* (CRC, Cleveland 1971)
2.17 E.I. Gordon: Proc. IEEE *54*, 1391 (1966)
2.18 T.G. Giallorenzi, A.F. Milton: J. Appl. Phys. *45*, 1762 (1974)
2.19 Y. Ohmachi: J. Appl. Phys. *44*, 3928 (1973)
2.20 H. Le Gall: "Magneto.Optic Effects and Materials" in *Photonics*, ed. by M. Balkanski, P. Lallemand (Gauthier-Villars, Paris 1975)
2.21 R.E. Collin: *Foundations for Microwave Engineers* (McGraw-Hill, New York 1966)
2.22 P.K. Tien, R.T. Martin, R. Wolfe, R.C. Le Graw, S.L. Blank: Appl. Phys. Lett. *21*, 397 (1972)
2.23 C.E. Land, P.D. Thacher, G.H. Heartling: "Electro-Optic Ceramics" in *Applied Solid State Science*, Vol.4 (Academic, New York 1974)
2.24 W.Z. Franz: Phys. Rev. *13*A, 484 (1958)
2.25 L.V. Keldysh: Zh. Eksp. Teor. Fiz. *7*, 788 (1958)
2.26 W. Känzig: Phys. Rev. *98*, 549 (1955)
2.27 R.H. Kingston: *Detection of Optical and Infrared Radiation*, Springer Series in Optical Sciences, Vol.10 (Springer, Berlin, Heidelberg, New York 1978)
2.28 R.J. Keyes (ed.): *Optical and Infrared Detectors*, 2nd ed., Topics in Applied Physics, Vol.19 (Springer, Berlin, Heidelberg, New York 1980)
2.29 H. Kressel (ed.): *Semiconductor Devices for Optical Communication*, Topics in Applied Physics, Vol.39 (Springer, Berlin, Heidelberg, New York 1979)
3.1 M. Born, E. Wolf: *Principles of Optics* (Pergamon, Oxford, New York 1975)
3.2 T. Tamir: Optik *36*, 209-232 (1972)
3.3 T. Tamir: Optik *37*, 204-228 (1973)
3.4 T. Tamir: Optik *38*, 269-297 (1973)
3.5 F. Goos, H. Hänchen: Ann. Phys. (Lpz.) *6*, 333 (1947)
3.6 H.K.V. Lotsch: Optik *32*, 116-137, 189-204 (1970)
3.7 H.K.V. Lotsch: Optik *32*, 299-319, 553-569 (1971)
3.8 H.K.V. Lotsch: J. Opt. Soc. Am. *58*, 551-561 (1968)
3.9 I. Newton: *Optics* (Dover, New York 1952) reproduction
3.10 B. Horowitz, T. Tamir: J. Opt. Soc. Am. *61*, 586 (1971)
3.11 A.I. Mahan, C.V. Bitterli: Appl. Opt. *17*, 509 (1978)
3.12 A.W. Snyder, I.D. Love: IEEE Trans. MTT-*23*, 134 (1975)
3.13 I.D. Love, A.W. Snyder: J. Opt. Soc. Am. *65*, 1072 (1975)
3.14 I.D. Love, A.W. Snyder: J. Opt. Soc. Am. *65*, 1241 (1975)
3.15 L.M. Brekhovskikh: *Waves in Layered Media* (Academic, New York 1960)
3.16 S. Kozaki, Y. Mushiake: J. Appl. Phys. *46*, 4098 (1975)
3.17 S. Kozaki, K. Kimura: J. Opt. Soc. Am. *66*, 63 (1976)
3.18 S. Kozaki, K. Kimura: Trans. IECE *59-B*, 238 (1976) (in Japanese)
3.19 H. Kogelnik: *Proc. Symp. Modern Optics*, ed. by I. Fox (Polytechnic Press, New York 1967)
3.20 H. Kogelnik: Bell Syst. Tech. J. *48*, 2909 (1969)
3.21 V. Shah, T. Tamir: Opt. Commun. *23*, 113 (1977)
3.22 R.I. Collier, Ch.B. Burckhardt, L.H. Lin: *Optical Holography* (Academic, New York 1971)
3.23 Ju.P. Udojev, V.A. Ovchinnikov: Opt. Spectroscopy *45*, 363 (1978)
3.24 J.W. Goodman: *Introduction to Fourier Optics* (McGraw-Hill, New York 1968)
3.25 H.K.V. Lotsch: Optik *27*, 339 (1968)
3.26 H. Kogelnik: "Theory of Dielectric Waveguides" in *Integrated Optics* 2nd ed., ed. by T. Tamir, Topics in Applied Physics, Vol.7 (Springer, Berlin, Heidelberg, New York 1979)

3.27 M.K. Barnoski: "Planar Waveguides - One Dimensional Case" in *Introduction to Integrated Optics*, ed. by M.K. Barnoski (Plenum, New York 1974)
3.28 D. Marcuse: *Light Transmission Optics* (Van Nostrand-Reinold, New York 1971)
3.29 H.-G. Unger: *Planar Optical Waveguides and Fibers* (Clarendon, Oxford 1977)
3.30 P.K. Tien: Rev. Mod. Phys. *49*, 361 (1977)
3.31 N. Fröman, P.O. Fröman: *JWKB Approximation* (North-Holland, Amsterdam 1965)
3.32 L. Landau, E. Lifschitz: *Electrodynamics of Continuous Media*, Theoretical Physics, Vol.8 (Mir, Moscow 1969)
3.33 M. Abramovitch, I.A. Stegun: *Handbook of Mathematical Functions* (Dover, New York 1970)
3.34 D. Hondros, P. Debye: Ann. Phys. *32*, 465-476 (1910)
3.35 N.S. Kapany, J.J. Burke: *Optical Waveguides* (Academic, New York 1972)
3.36 D. Gloge: IEEE Trans. MTT-*23*, 106 (1975)
3.37 G. Arfken: *Mathematical Methods for Physicists* (Academic, New York 1970)
3.38 A.B. Sharma, J. Saijonamaa, S.J. Halme: Electron. Lett. *16*, 557 (1980)
3.39 D. Gloge, E.A.J. Marcatili: Bell Syst. Tech. J. *52*, 1563 (1973)
3.40 I.S. Gradshteyn, I.M. Ryzhik: *Tables of Integrals, Series, and Products* (Academic, New York 1965)
3.41 J.J. Ramskov.Hansen: Appl. Opt. *17*, 2831 (1978)
3.42 J.W. Goodman: J. Opt. Soc. Am. *66*, 1145 (1976)
3.43 J.W. Goodman: "Statistical Properties of Laser Speckle Patterns" in *Laser Speckle and Related Phenomena*, ed. by J.C. Dainty, Topics in Applied Physics, Vol.9 (Springer, Berlin, Heidelberg, New York 1975)
3.44 B. Crosignani, B. Daino, P. DiPorto: Opt. Commun. *11*, 178 (1974)
3.45 B. Crosignani, B. Daino: J. Opt. Soc. Am. *66*, 1312 (1976)
3.46 M. Imai, T. Asakura: Optik *48*, 335 (1977)
4.1 K.C. Kao, G.A. Hockham: Proc. IEE (London) *113*, 1151 (1966)
4.2 S.E. Miller, T. Li, E.A.J. Marcatili: Proc. IEEE *61*, 1703 (1973)
4.3 R.D. Maurer: Proc. IEEE *61*, 452 (1973)
4.4 N.S. Kapany: *Fiber Optics* (Academic, New York 1967)
4.5 J.E. Midwinter: *Optical Fibers for Transmission* (Wiley, New York 1979) pp.162-196
4.6 K. Koizumi, Y. Ikeda, I. Kitano, M. Furukawa, T. Sumimoto: Appl. Opt. *13*, 255 (1974)
4.7 K.B. Chan, P.J.B. Clarricoats, R.B. Dyott, G.R. Newns, M.A. Savva: Electron. Lett. *6*, 748 (1970)
4.8 K.J. Beales, C.R. Day, W.J. Duncan, A.G. Dunn, G.R. Newns, S. Partington: "Preparation of Low Loss Graded-Index and High N.A. Step-Index Compound Glass Fiber by the Double Crucible Technique", in *Abstracts*, American Ceramic Society Annual Meeting, Cincinnati (May 1979) abstract 38-G-79, p.378
4.9 M.G. Blankenship, D.B. Keck, P.S. Levin, W.F. Love, R. Olshansky, A. Sarkar, P.C. Schultz, K.D. Sheth, R.W. Siegfried: "High Phosphorus-Containing P_2O_5-GeO_2-SiO_2 Optical Waveguide", Post-dead-line paper, OSA Topical Meeting on Optical Fiber Communication, Washington DC (1979)
4.10 D. Kuppers, H. Lydtin: "Preparation Methods for Optical Fibers Applied in Philips Research" Proc. International Conference on Integrated Optics and Optical Fiber Communication, Tokyo (1977) (IECE, Tokyo 1977)
4.11 P.C. Schultz: Appl. Opt. *18*, 3684 (1979)
4.12 D.B. Keck, R. Boullie: Opt. Commun. *25*, 43 (1978)
4.13 H. Murata (ed.): "The Review of Recent Development of Optical Fiber and Cable Development in Japan", Report TI 79027, Furukawa Electric Co. Ltd., Tokyo, Japan (1979)

4.14 H. Kroemer: Proc. IEEE *51*, 1782 (1963)
4.15 Zh.I. Alferov: Fiz. i Tekhnika Polyprovodnikov *1*, 436 (1967)
4.16 B.L. Sharma, R.K. Purohit: *Semiconductor Heterojunctions* (Pergamon, Oxford, 0000 1974)
4.17 S.E. Miller, A.G. Chynoweth (eds.): *Optical Fiber Telecommunications* (Academic, New York 1979)
4.18 H. Kressel (ed.): *Semiconductor Devices for Optical Communication*, Topics in Applied Physics, Vol.39 (Springer, Berlin, Heidelberg, New York 1979)
4.19 H. Kressel, J.K. Butler: *Semiconductor Lasers and Heterojunction LEDS* (Academic, New York 1977)
4.20 H.C. Cassey Jr., M.B. Panish: *Heterostructure Lasers*, (Academic, New York 1978)
4.21 E. Garmire: "Semiconductor Components for Monolithic Applications" in *Integrated Optics*, 2nd ed., ed. by T. Tamir, Topics in Applied Physics, Vol.7 (Springer, Berlin, Heidelberg, New York 1979)
4.22 J.C. Dyment, L.A. D'Asaro, J.C. North, B.I. Miller, J.E. Ripper: Proc. IEEE *60*, 726 (1972)
4.23 T. Ozeki, T. Ito: IEEE J. QE-*9*, 1098 (1973)
4.24 T.P. Lee, R.M. Derosier: Proc. IEEE *62*, 1176 (1974)
4.25 G. Arnold, P. Russer: Appl. Phys. *14*, 255 (1977)
4.26 H. Kogelnik, C.V. Shank: Appl. Phys. Lett. *18*, 152 (1971)
4.27 H. Kogelnik, C.V. Shank: J. Appl. Phys. *43*, 2327 (1972)
4.28 M. Nakamura, A. Yariv, H.W. Yen, S. Somekh, H.L. Garvin: Appl. Phys. Lett. *22*, 315 (1973)
4.29 M. Nakamura, H.W. Yen, A. Yariv, E. Garmire, S. Somekh, H.L. Garvin: Appl. Phys. Lett. *23*, 224 (1973)
4.30 M. Nakamura, K. Aiki, J. Umeda, A. Yariv, H.W. Yen, T. Morikawa: Appl. Phys. Lett. *25*, 487 (1974)
4.31 C.V. Shank, R.V. Schmidt: Appl. Phys. Lett. *25*, 200 (1974)
4.32 D.R. Scifres, R.D. Burnham, W. Streifer: Appl. Phys. Lett. *25*, 203 (1974)
4.33 H.C. Casey Jr., S. Somekh, M. Ilegems: Appl. Phys. Lett. *27*. 142 (1975)
4.34 M. Nakamura, K. Aiki, J. Umeda, A. Yariv: Appl. Phys. Lett. *27*, 403 (1975)
4.35 C.L. Tang: "Laser Sources in Integrated Optics" in *Introduction to Integrated Optics*, ed. by M.K. Barnoski (Plenum, New York 1974)
4.36 P.K. Tien: Rev. Mod. Phys. *49*, 361 (1977)
4.37 A. Yariv, M. Nakamura: IEEE J. QE-*13*, 233 (1977)
4.38 J. Stone, C.A. Burrus: Appl. Opt. *13*, 1256 (1974)
4.39 C.A. Burrus, J. Stone: J. Appl. Phys. *49*, 2281 (1978)
4.40 J. Stone, C.A. Burrus: Fiber and Integrated Optics *2*, 19 (1979)
4.41 I.P. Kaminow, J.R. Carruthers: Appl. Phys. Lett. *22*, 326 (1973)
4.42 J. Noda, N. Uchida, S. Saito, T. Saku, M. Minakata: Appl. Phys. Lett. *27*, 19 (1975)
4.43 R.V. Schmidt, H. Kogelnik: Appl. Phys. Lett. *26*, 503 (1976)
4.44 G.L. Tangonan, M.K. Barnoski, J.F. Lotspeich, A. Lee: Appl. Phys. Lett. *30*, 238 (1977)
4.45 J.M. Hammer, W. Phillips: Appl. Phys. Lett. *24*, 545 (1974)
4.46 A. Okada: Ferroelectrics, *14*, 739 (1976)
4.47 S. Miyazawa, S. Fushimi, S. Kondo: Appl. Phys. Lett. *26*, 8 (1975)
4.48 S. Kondo, S. Miyazawa, S. Fushimi, K. Sugii: Appl. Phys. Lett. *26*, 489 (1975)
4.49 S. Miyazawa, K. Sugii, N. Uchida: J. Appl. Phys. *46*, 2223 (1975)
4.50 B.J. Curtis, H.R. Brunner: Mat. Res. Bull. *10*, 515 (1975)
4.51 I.P. Kaminow, L.W. Stulz, E.H. Turner: Appl. Phys. Lett. *27*, 555 (1975)
4.52 N. Uchida: Opt. Quant. Electron. *9*, 1 (1977)

4.53 W. Martin: Appl. Phys. Lett. *26*, 562 (1975)
4.54 J.M. Hammer, D.J. Channin, M.T. Duffy, C.C. Neil: IEEE J. QE-*11*, 138 (1975)
4.55 P.K. Tien, S. Riva-Sanseverino, A.A. Ballman: Appl. Phys. Lett. *25*, 563 (1974)
4.56 J.C. An, Y. Cho, Y. Matsuo: IEEE J. QE-*13*, 206 (1977)
4.57 I.P. Kaminow, L.W. Stulz: IEEE J. QE-*11*, 633 (1975)
4.58 C.S. Tsai, P. Sauner: Appl. Phys. Lett. *27*, 248 (1975)
4.59 T.P. Lee, T. Li: "Photodetectors" in *Optical Fiber Telecommunications*, ed. by S.E. Miller, A.G. Chynoweth (Academic, New York 1979)
4.60 D.P. Schinke, R.G. Smith, A.R. Hartman: "Photodetectors" in *Semiconductor Devices for Optical Communication*, ed. by H. Kressel, Topics in Applied Physics, Vol.39 (Springer, Berlin, Heidelberg, New York 1979)
4.61 R.J. Keyes (ed.): *Optical and Infrared Detectors*, 2nd ed., Topics in Applied Physics, Vol.19 (Springer, Berlin, Heidelberg, New York 1980)
4.62 H. Melchior, A.R. Hartman, D.P. Schinke, T.E. Seidel: Bell Syst. Tech. J. *57*, 1791 (1978)
4.63 H. Melchior, M.B. Fisher, F.R. Arams: Proc. IEEE *58*, 1466 (1970)
4.64 H. Melchior: "Demodulation and Photodetection Techniques" in *Laser Handbook*, ed. by F.T. Arechi, E.D. Schulz-Dubois (Elsevier, North-Holland, New York 1972)
4.65 R.K. Willardson, A.C. Beer (eds.): *Semiconductors and Semimetals*, Vols. 5 and 12 (Academic, New York 1970, 1977)
4.66 T. Kaneda, H. Takanashi, H. Matsumoto, T. Yamaoka: J. Appl. Phys. *47*, 4960 (1976)
4.67 P.P. Webb, R.J. McIntyre, J. Conradi: RCA Rev. *35*, 234 (1974)
4.68 R.B. Emmons: J. Appl. Phys. *38*, 3705 (1967)
4.69 D. Wolf (ed.): *Noise in Physical Systems*, Springer Series in Electrophysics, Vol.2 (Springer, Berlin, Heidelberg, New York 1978)
4.70 R.J. McIntyre: IEEE Trans. ED-*19*, 703 (1972)
4.71 D.L. Bisbee: Bell Syst. Tech. J. *50*, 3153 (1971)
4.72 Y. Kohanzadeh: Appl. Opt. *15*, 793 (1976)
4.73 K. Sakamoto, T. Miyajiri, H. Kakuzen, M. Hirai, N. Uchida: Proc. 4th Eur. Conf. Opt. Fib. Comm., Genoa (1978)
4.74 T. Arai, O. Watanabe, K. Inada, Y. Katsuyama: Proc. 5th Eur. Conf. Opt. Fib. Comm., Amsterdam (1979)
4.75 M. Hirai, W. Naoya: Electron. Lett. *13*, 123 (1977)
4.76 I. Hatakeyama, H. Tsuchiya: Appl. Opt. *17*, 1959 (1978)
4.77 I. Hatakeyama, H. Tsuchiya: IEEE J. QE-*14*, 614 (1978)
4.78 C.G. Someda: Bell Syst. Tech. J. *52*, 583 (1973)
4.79 R.M. Derosier, J. Stone: Bell Syst. Tech. J. *52*, 1229 (1973)
4.80 C.M. Miller: Bell Syst. Tech. J. *54*, 1215 (1975)
4.81 R.D. Standley: Bell Syst. Tech. J. *53*, 1183 (1974)
4.82 C.M. Schroeder: Bell Syst. Tech. J. *57*, 91 (1978)
4.83 D. Gloge, A.H. Cherin, C.M. Miller, P.W. Smith: "Fiber Splicing" in *Optical Fiber Telecommunications*, ed. by S.E. Miller, A.G. Chynoweth (Academic, New York 1979)
4.84 J. Cook, P.K. Runge: "Optical Fiber Connectors" in *Optical Fiber Telecommunications*, ed. by S.E. Miller, A.G. Chynoweth (Academic, New York 1979)
4.85 M. Börner, D. Gruchmann, J. Guttman, O. Krumpholz, W. Löfler: Arch. Elek. Übertrag. *26*, 288 (1972)
4.86 N. Kurochi, A. Ushirogawa, Y. Morimoto, M. Shimozono: Proc. 3rd Eur. Conf. Opt. Fib. Comm., München (1977)
4.87 V. Vucins: Proc. 3rd Eur. Conf. Opt. Fib. Comm. München (1977)
4.88 G.D. Khoe, H.W.W. Smulders, A.J.J. Franken: Proc. 5th Eur. Conf. Opt. Fib. Comm., Amsterdam (1979)

4.89 J.C. North, J.H. Stwart: Proc. 5th Eur. Conf. Opt. Fib. Comm., Amsterdam (1979)
4.90 D. Marcuse: Bell Syst. Tech. J. *49*, 1695 (1970)
4.91 A. Cardama, E.T. Kornhauser: IEEE Trans. MTT-*23*, 162 (1975)
4.92 H.-G. Unger: *Planar Optical Waveguides and Fibers* (Clarendon, Oxford 1977) Chap.8
4.93 D. Marcuse: *Light Transmission Optics* (Van Nostrand-Reinold, New York 1971)
4.94 M.C. Hudson: Appl. Opt. *13*, 1029 (1974)
4.95 M.K. Barnoski (ed.): *Fundamentals of Optical Fiber Communications* (Academic, New York 1976) Chap.3
4.96 D. Kato: J. Appl. Phys. *44*, 2756 (1973)
4.97 M. Abe, I. Umebu, O. Hasegawa, S. Yamokoshi, T. Yamaoka, T. Kotani, H. Okada, H. Takanashi: IEEE Trans. ED-*24*, 991 (1977)
4.98 C.P. Basola, G. Chiaretti: Proc. 4th Eur. Conf. Opt. Fib. Comm., Genoa (1978)
4.99 T. Ozeki, B.S. Kawasaki: Electron. Lett. *12*, 607 (1976)
5.1 M.K. Barnoski (ed.): *Fundamentals of Optical Fiber Communications* (Academic, New York 1976)
5.2 J.E. Midwinter, J.R. Stern: IEEE Trans. COM-*26*, 1015 (1978)
5.3 S. Geckeler: Appl. Opt. *18*, 2192 (1979)
5.4 D. Gloge, P.W. Smith, D.L. Bisbee, E.L. Chinnock: Bell Syst. Tech. J. *52*, 1579 (1973)
5.5 J.E. Midwinter: *Optical Fibers for Transmission* (Wiley, New York 1979)
5.6 T. Miya, Y. Terunuma, T. Hosaka, T. Miyashita: Electron. Lett. *15*, 106 (1979)
5.7 T. Tanifuji, T. Horiguchi, M. Tokuda: Electron. Lett. *15*, 203 (1979)
5.8 S.D. Personick: Bell Syst. Tech. J. *56*, 355 (1977)
5.9 M.K. Barnoski, M.D. Rourke, S.M. Jensen, R.T. Melville: Appl. Opt. *16*, 2375 (1977)
5.10 A.B. Carlson: *Communication Systems* (McGraw-Hill Kogakusha, Tokyo 1968)
5.11 J.R. Andrews: Rev. Sci. Instrum. *45*, 22 (1974)
5.12 W.D. Roehr (ed.): *Switching Transistor Handbook* (Motorola, Inc., 1967)
5.13 S.D. Personick: Bell Syst. Tech. J. *52*, 1175 (1973)
5.14 M. Maeda, K. Nagano, Y. Minai, M. Tanaka: Appl. Opt. *17*, 651 (1978)
5.15 C. Boisrobert, A. Cozannet, C. Vassalo: IEEE Trans. IM-*25*, 294 (1976)
5.16 S.D. Personick, W.M. Hubbard, W.S. Holden: Appl. Opt. *13*, 266 (1974)
5.17 E. Nicolaisen, J.J. Ramskov-Hansen: Proc. 4th Eur. Conf. Opt. Fib. Comm., Genoa (1978)
5.18 A.B. Sharma, S.J. Halme: Appl. Opt. *18*, 1877 (1979)
5.19 W.E. Martin: Appl. Opt. *13*, 2112 (1974)
5.20 D. Gloge, E.A.J. Marcatili: Bell Syst. Tech. J. *52*, 1563 (1973)
5.21 T. Okoshi, K. Hotate: Appl. Opt. *15*, 11 (1976)
5.22 M. Ikeda, M. Tateda, H. Yoshikiyo: Appl. Opt. *14*, 814 (1975)
5.23 W. Eickhoff, E. Weidel: Opt. Quant. Electron. *7*, 109 (1975)
5.24 E. Brinkmeyer: Appl. Opt. *17*, 14 (1978)
5.25 W.J. Stewart: Proc. 4th Eur. Conf. Opt. Fib. Comm., Genoa (1978)
5.26 F.T. Stone: Appl. Opt. *16*, 2738 (1977)
5.27 M. Tateda: Appl. Opt. *17*, 475 (1978)
5.28 M. Ito, M. Okada, T. Miya: Rev. Electr. Comm. Labs. (Japan) *26*, 73 (1978)
5.29 H.M. Presby, D. Marcuse, H.W. Astle: Appl. Opt. *17*, 2209 (1978)
5.30 M.J. Adams, D.N. Payne, F.M.E. Sladen: Opt. Commun. *17*, 204 (1976)
5.31 M.J. Adams, D.N. Payne, F.M.E. Sladen: Electron. Lett. *12*, 281 (1976)
5.32 I.S. Gradshteyn, I.W. Ryzhik: *Tables of Integrals, Series, and Products* (Academic, New York 1965)
5.33 R. Olshansky, S.M. Oaks: Proc. 4th Eur. Conf. Opt. Fib. Comm., Genoa (1978)

5.34 J. Saijonmaa, A.B. Sharma, S.J. Halme: Appl. Opt. *19*, 2442 (1980)
6.1 P. Bylanski, D.G.W. Ingram: *Digital Transmission Systems*, IEE Telecommunications Series 4 (Peter Peregrinus, Stevenage 1976)
6.2 Posts and Telecommunications of Finland: *Regulations for Telephone Networks* (Helsinki 1979)
6.3 U. Haller, W. Herold, H. Ohnsorge: Appl. Phys. *17*, 115 (1978)
6.4 B.C. De Loach, R.C. Miller, S. Kaufmann: Bell Syst. Tech. J. *57*, 3309 (1978)
6.5 R.C. Miller, R.B. Lawry: Bell Syst. Tech. J. *58*, 1735 (1979)
6.6 L.W. Ellis: Electrical Communications *50*, 1-19 (1975)
6.7 M. Born, E. Wolf: *Principles of Optics* (Pergamon, Oxford 1975)
6.8 Y. Takasaki, M. Tanaka, N. Malda, K. Yamashita, K. Nagano: IEEE COM-*24*, 404 (1976)
6.9 J.E. Savage: Bell Syst. Tech. J. *48*, 449 (1969)
6.10 M. Hecht, A. Guida: Proc. IEEE *57*, 1316 (1969)
6.11 R. Petrovic: Electron. Lett. *14*, 806 (1978)
6.12 A. van der Ziel: Proc. IEEE *58*, 1178 (1970)
6.13 S.D. Personick: Bell Syst. Tech. J. *52*, 843 (1973)
6.14 J.E. Goell: Bell Syst. Tech. J. *53*, 629 (1974)
6.15 A. Papoulis: *Probability, Random Variables, and Stochastic Processes* (McGraw-Hill Kogakusha, Tokyo 1975)
6.16 H.L. Van Trees: *Detection, Estimation, and Modulation Theory*, Pt.1 (Wiley, New York 1968)
6.17 M. Abramowitz, I.A. Stegun: *Handbook of Mathematical Functions* (Dover, New York 1970)
6.18 S. Karp, E.L. O'Neill, R.L. Gagliardi: Proc. IEEE *58*, 1611 (1970)
6.19 R.W. Lucky, J. Salz, E.J. Weldon: *Principles of Data Communication* (McGraw-Hill, New York 1968)
6.20 A.J. Viterbi (ed.): *Advances in Communications Systems* (Academic, New York 1975)
6.21 A.V. Tenne-Sens, D.C. Johnson: IEEE Trans. CATV-*3*, 145 (1978)
6.22 E.H. Hara, T. Ozeki: IEEE Trans. CATV-*2*, 18 (1977)
6.23 D. Chan, T.M. Yuen: IEEE Trans. COM-*25*, 680 (1977)
6.24 T. Nakahara, H. Kumamaru, T. Takeuchi: IEEE Trans. COM-*26*, 955 (1978)
6.25 R.L. Freeman: *Telecommunication Transmission Handbook* (Wiley, New York 1975)
6.26 R.E. Epworth: Proc. 4th Eur. Conf. Opt. Fib. Comm., Genoa (1978)
6.27 C.C. Timmermann: IEEE Trans. BC-*23*, 12 (1977)
6.28 J.M. Wozencraft, I.M. Jacobs: *Principles of Communication Engineering* (Wiley, New York 1965)
6.29 M.C. Hudson, F.L. Thiel: Appl. Opt. *13*, 2540 (1974)
6.30 T. Ozeki, B.S. Kawasaki: Electron. Lett. *12*, 151 (1976)
6.31 B. Kincaid: Appl. Opt. *16*, 2355 (1977)
6.32 B.S. Kawasaki, K.O. Hill: Appl. Opt. *16*, 1794 (1977)
6.33 D.H. McMahon, R.L. Gravel: Appl. Opt. *16*, 501 (1977)
6.34 G.B. Hocker: Opt. Lett. *1*, 124 (1977)
6.35 E.G. Rawson, R.M. Metcalfe: IEEE Trans. COM-*26*, 983 (1978)
6.36 CCITT Orange Book VIII. 1 (ITU, Geneva 1977)
6.37 IEEE Std. 488 (1978)
6.38 D.D. Clark, K.T. Pogran, D.P. Reed: Proc. IEEE *66*, 1497 (1978)
6.39 Opto-Electronic Designer's Catalog (Hewlett-Packard, Palo Alto, CA 1979)
6.40 W.W. Brown, D.C. Hanson, T. Hornak, S.F. Garvey: IEEE Trans. COM-*26*, 976 (1978)
6.41 D.C. Hanson, W. Brown, S. Carvey, G. Girot, E. Heldt: IEEE Trans. COM-*26*, 1068 (1978)
6.42 G.B. Hocker: Appl. Opt. *18*, 1445 (1979)
6.43 J.A. Bucaro, T.R. Hickman: Appl. Opt. *18*, 938 (1979)

6.44 J.A. Bucaro, H.D. Dardy, E.F. Carome: Appl. Opt. *16*, 1761 (1977)
6.45 B. Culshaw, D.E.N. Davies, S.A. Kingsley: Electron. Lett. *13*, 761 (1977)
6.46 V. Vali, R.W. Shorthill, M.F. Berg: Appl. Opt. *16*, 2605 (1977)
6.47 S.-C. Liu, T.G. Giallorenzi: Appl. Opt. *18*, 915 (1979)
6.48 H. Ishio: Proc. 5th Eur. Conf. Opt. Comm., (Post deadline), Amsterdam (1979)
6.49 M. Young: *Optics and Lasers*, Springer Series in Optical Sciences, Vol.5 (Springer, Berlin, Heidelberg, New York 1977)
6.50 K. Hashimoto, K. Nosu: Proc. 5th Eur. Conf. Opt. Comm., Amsterdam (1979)

Subject Index